Animal Welfare

Series Editor

Professor Clive Phillips
Foundation Chair of Animal Welfare
Centre for Animal Welfare and Ethics
School of Veterinary Science
University of Queensland
Gatton 4343, QLD
Australia

For further volumes:
http://www.springer.com/series/5675

Edward N. Eadie

Understanding Animal Welfare

An Integrated Approach

Springer

Edward N. Eadie

ISSN 1572-7408
ISBN 978-3-642-30576-4 ISBN 978-3-642-30577-1 (eBook)
DOI 10.1007/978-3-642-30577-1
Springer Heidelberg New York Dordrecht London

Library of Congress Control Number: 2012945292

Printed on acid-free paper

Springer is part of Springer Science+Business Media (www.springer.com)

In Memoriam: Dr Edward N. Eadie (1935–2012)

Dr Edward Eadie was born in 1935 to Norman, a medical doctor who served on the Burma railway and was a PoW in the Second World War, and Eileen Eadie of Melbourne. His first degree was in mathematics and physics from the University of Melbourne, after which he joined the Australian Bureau of Mineral Resources as a geophysicist. He explored the Savage River in Tasmania for iron ore deposits, which resulted in an industry sending two million tons of iron ore concentrate to Japan and enabled him to write a thesis for a Master's degree from the University of Melbourne. After periods as a mining investment analyst and Chief Geophysicist with McPhar Geophysics in Canada, he went to Oxford University and obtained a doctorate in integrated approaches to mining exploration. This was perhaps a foretaste of the integrated approaches to animal protection that he was to advocate later, bringing together scientists, politicians, lawyers, and animal protection organisations to address injustices to animals. However, his efforts towards animal protection were to remain dormant whilst he worked in international trade in minerals, for which he was awarded a Ph.D. by the University of Adelaide, and maritime law, for which he was awarded Master's degrees from the University of London, the Australian Maritime College, and the University of Wollongong. He devoted the remaining 10 years of his life to understanding and writing about animal suffering. He championed the role of law and education in alleviating, in his words, 'the terrible situation in which many animals in various circumstances find themselves throughout the world'. After travelling in more than 50 countries, he had no doubt witnessed many scenes that made him determined to concentrate on animal welfare issues in his later years. He wrote a thesis on Animal Suffering and the Law for the Australian National University, which was also published as a book by Seaview Press. He also studied educational themes and contexts for reducing animal suffering, which formed the basis for a book 'Education in Animal Welfare', published as part of this Animal Welfare Series by Springer, and earned him a Master's degree from the University of Sydney. Finally he joined the Centre for Animal Welfare and Ethics at the University of Queensland to work on a selection of modern books to improve animal welfare, published in this volume and for which he was awarded a posthumous Master's degree by the University of Queensland. To borrow a phrase gaining credibility in animal welfare studies, Ted's life was truly a 'life worth living'.

The royalties from this book will be donated to the Ted Eadie Memorial Scholarship, which will support the work of the Centre for Animal Welfare and Ethics in alleviating animal suffering.

Animal Welfare Series Preface

Animal welfare is attracting increasing interest worldwide, especially in developed countries where the knowledge and resources are available to at least potentially provide better management systems for farm animals, as well as companion, zoo, and laboratory animals. The key requirements for adequate food, water, a suitable environment, companionship, and health are important for animals kept for all of these purposes.

There has been increased attention given to farm animal welfare in the West in recent years. This derives largely from the fact that the relentless pursuit of financial reward and efficiency, to satisfy market demands, has led to the development of intensive animal production systems that challenge the conscience of many consumers in this part of the world. Livestock are the world's biggest land users (FAO 2002), and the farmed animal population is increasing rapidly to meet the needs of an expanding human population. This results in a tendency to allocate fewer resources to each animal and to value individual animals less, for example, in the case of farmed poultry where flocks of over 20,000 birds are not uncommon. In these circumstances, the importance of each individual's welfare is diminished.

In developing countries, human survival is still a daily uncertainty, so that provision for animal welfare has to be balanced against human welfare. Animal welfare is usually a priority only if it supports the output of the animal, be it food, work, clothing, sport, or companionship. However, in many situations, the welfare of animals is synonymous with the welfare of the humans that look after them, because happy, healthy animals will be able to assist humans best in their struggle for survival. In principle, the welfare needs of both humans and animals can be provided for, in both developing and developed countries, if resources are properly husbanded. In reality, the inequitable division of the world's riches creates physical and psychological poverty for humans and animals alike in many parts of the world.

Increased attention to welfare issues is just as evident for companion, laboratory, wild, and zoo animals. Of increasing importance is the ethical management of breeding programmes, since genetic manipulation is now technically advanced, but there is less public tolerance of the breeding of extreme animals if it comes at the expense of animal welfare. The quest for producing novel genotypes has fascinated breeders for centuries. Dog and cat breeders have produced a variety of deformities that have adverse effects on their welfare, but nowadays the breeders are just as active in the laboratory, where the mouse is genetically manipulated with equally profound effects.

The intimate connection between animals and humans that was once so essential for good animal welfare is rare nowadays, having been superseded by technologically efficient production systems where animals on farms and in laboratories are tended by increasingly few humans in the drive to enhance labour efficiency. With today's busy lifestyles, companion animals too may suffer from reduced contact with humans, although their value in providing companionship, particularly for certain groups such as the elderly, is increasingly recognised. Consumers also rarely have any contact with the animals that are kept for their benefit.

In this estranged, efficient world, people struggle to find the moral imperatives to determine the level of welfare that they should afford to animals within their charge. Some, and in particular many companion animal owners, strive for what they believe to be the highest levels of welfare provision, while others, deliberately or through ignorance, keep animals in impoverished conditions in which their health and well-being can be extremely poor. Today's multiple moral codes for animal care and use are derived from a broad range of cultural influences, including media reports of animal abuse, guidelines on ethical consumption, and campaigning and lobbying groups.

This series has been designed to contribute towards a culture of respect for animals and their welfare by producing learned treatises about the provision for the welfare of the animal species that are managed and cared for by humans. The early species-focused books were not detailed management blueprints; rather they described and considered the major welfare concerns, often with reference to the behaviour of the wild progenitors of the managed animals. Welfare was specifically focused on animals' needs, concentrating on nutrition, behaviour, reproduction, and the physical and social environment. Economic effects of animal welfare provision were also considered where relevant, as were key areas where further research is required.

In this volume, the series again departs from the species focus to address our literature on animal welfare over approximately the past 50 years. Dr Ted Eadie had a lifetime of achievement and sadly passed from us in June, 2012. He remained productive as an academic right up until the end. In his early years, he worked mainly in the minerals and maritime industries, gaining doctorates in science, commerce, and law from Oxford, Adelaide, and the Australian National University. Later in life, he devoted his life to alleviating the suffering of animals. His first two books addressed animal welfare in relation to the law (Animal Suffering and the Law, 2009, Seaview Press) and education (Education for Animal Welfare, 2011, Springer), both receiving excellent reviews. In this latest, and final, work Ted addresses the challenging topic of synthesising the major writings about animal welfare. He focuses on key texts in eight central animal welfare areas: animal experimentation, intensive farming, immorality and animal ethics, speciesism, animal welfare science, practical action to reform animal ethics, animal protection politics, and legislation. In each area, he outlines the contribution provided by a major text in the field, other books with potential for influence, as well as detailed consideration of each area of interest. Ted provides an historical perspective of the development of animal welfare interests. His strong belief in the causes espoused in each of these fields of animal welfare, his determination that animals should be given better treatment, and his unfailing respect for those that pioneered the discipline of animal welfare provide a book that is a delight to read, full of hope, and a positive attitude to these enormous challenges. Ted remained confident that we could change the way animals are treated if the issues were more widely understood. His life and this final book are truly worth celebrating.

Gatton, Queensland, Australia Clive Phillips

Reference

Food and Agriculture Organisation (2002) http://www.fao.org/ag/aga/index_en.htm

Series Editor's notes

The term animal or creature refers to nonhuman animals. While every attempt to use non-gendered language in reference to human-animal interactions, this has been retained if it is quoted from another source, which were often written before the inequalities of gendered language were commonly recognised.

Abstracts for each chapter were written by the Editor. Helpful comments were received from Dr Siobhan O'Sullivan of the University of Melbourne and Professor Kevin Stafford of Massey University.

Acknowledgements

The author wishes to thank the many people who have contributed in various ways towards the production of this book. Special thanks are due to Professor Clive Phillips, Director of the Centre for Animal Welfare and Ethics in the School of Veterinary Science at the University of Queensland and Editor of the Springer Animal Welfare Series, for his wonderful support and friendship throughout the research for and writing of the book; Mrs. Jillian Stevens of Adelaide for her dedication with the word processing; and Mrs. Joan Carter of Adelaide for her inspiration generally and interest in animal protection matters.

Contents

Abbreviations

ABC	Australian Broadcasting Corporation
APC	Animal Procedures Committee
BSE	Bovine Spongiform Encephalopathy
EC	European Community
ECVAM	European Centre for the Validation of Alternative Methods
EU	European Union
FAWC	Farm Animal Welfare Council
FRAME	Fund for the Replacement of Animals in Medical Experiments
HSUS	Humane Society of the United States
MAFF	Ministry of Agriculture, Fisheries and Food
MP	Member of Parliament
NFU	National Farmers Union
NSW	New South Wales
OIE	World Organisation for Animal Health
PETA	People for the Ethical Treatment of Animals
RCVS	Royal College of Veterinary Surgeons
RSPCA	Royal Society for the Prevention of Cruelty to Animals
SPCA	Society for the Prevention of Cruelty to Animals
SSPCA	Scottish Society for the Prevention of Cruelty to Animals
Three Rs	Replacement, reduction and refinement
UFAW	Universities Federation for Animal Welfare
UK	United Kingdom
US	United States
USA	United States of America
WSPA	World Society for the Protection of Animals
WTO	World Trade Organisation

Chapter 1
Introduction

Abstract This book focuses on writers' attempts to bring animal welfare issues to the attention of the public and relevant professionals. The major areas of interest are identified as intensive farming, animal experimentation, alleviation of animal suffering, human-animal relationships, animal welfare science, animal activism, animal politics, and animal law. Key texts written over the last 50 years are reviewed, and their contribution to knowledge and the development of animal welfare as a discipline are explored.

This book examines a selection of modern books published over the last 50 years or so that are relevant to improving the welfare of animals whose lives are affected by the control or actions of humans. The books were selected by Dr Eadie, with assistance from Professor Phillips, on the grounds of their perceived impact, with recent books necessarily excluded. The books selected have been chosen to cover important issues including experimental animals and intensive farming as well as a range of disciplines that include science, philosophy, history, activism, politics, and law. A separate chapter is devoted to each selected book, and the books selected are arranged in order of their date of publication. For each selected book, its contents are outlined then its influence and potential are examined. Moreover, for each book selected, some other related books are mentioned or discussed. Naturally, in the case of a selected book published many years ago, there is greater scope for examining its influence, whereas for a more recently published book, greater emphasis has to be placed on its potential for future influence. The number of books selected for detailed discussion is limited by the desire to deal with each in reasonable depth as well as cover certain specific issues or disciplines, and it needs to be emphasised that the choice of books is a personal one. Inevitably, some good books are neither selected nor even mentioned, but this should not be seen as detracting in any way from their value in making a useful contribution to knowledge and improving the welfare of animals. Indeed, advances in the creation of a better life for animals are achieved through the efforts of a great many dedicated people, both those who write books and those who contribute in other ways.

E.N. Eadie, *Understanding Animal Welfare*, Animal Welfare 13,
DOI 10.1007/978-3-642-30577-1_1, © Springer-Verlag Berlin Heidelberg 2012

Since the middle of last century, there have been major changes relevant to the relationship between animals and humans, particularly in Western countries, as observed by Rollin (2003, p. xiii), Webster (1995, p. 135), and Phillips (2011a, p. v). These changes included:

1. A large growth since the Second World War in both industrial agriculture and biomedical research that cause animal suffering but are not motivated by cruelty.
2. Since around 1950, there have been greater changes in livestock production than at any other time in the history of agriculture, with livestock farmers expanding production to meet unprecedented demand at reduced production cost by removing animals from the land and enclosing them in buildings.
3. Intensive farming practices resulted in less space being allocated to each animal so its individual welfare diminished.
4. During the latter half of the 1900s, there was a major increase in ethical sensitivity towards disenfranchised humans based on factors such as race and gender, and the obvious next step was for such sentiments to be extended to animals.
5. There was a huge decrease in the proportion of the human population working directly with animals, and animals became viewed as companions rather than as working animals or as a source of food.
6. The urban population became increasingly fascinated by animals as well as interested in their behaviour and treatment.
7. A number of extremely intelligent and articulate people, particularly philosophers and some scientists, saw the need for a new ethical perspective regarding the treatment of animals by humans and attempted to provide it.
8. A rapidly growing human population and its demands have resulted in increased competition with animals for land, and this has caused a significant amount of animal habitat destruction.

Thus, it is particularly appropriate to consider a selection of books published over the previous 50 years or so.

Only eight books have been selected for detailed consideration, although in the case of each some other related books are also mentioned or discussed. It is appropriate to outline here the rationale behind the choice of the primary books selected. The first three of these books are seminal works of great significance in relation to animal experimentation, intensive farming, and the immorality of inflicting suffering on animals. These books are *The Principles of Humane Experimental Technique* by William Russell and Rex Burch published in 1959 (Russell and Burch 1959) that proposed the *Three Rs* that are now widely recognised for eliminating or reducing the suffering of animals during experimentation; *Animal Machines: The New Factory Farming Industry* by Ruth Harrison published in 1964 (Harrison 1964) that exposed the suffering inflicted on animals in modern factory farming enterprises and resulted in the establishment in 1965 of the Brambell Committee in Britain from which the now well-known *Five Freedoms* for animals evolved; and *Animal Liberation: Towards an End of Man's Inhumanity to Animals* by Peter Singer published in 1975 (Singer 1975) that showed the moral indefensibility of inflicting suffering on animals and exposed existing discrimination by humans against animals known as *speciesism* both in animal experimentation or testing and in factory farming situations.

The next two books selected are *Animal Revolution: Changing Attitudes Towards Speciesism* by Richard Ryder published in 1989 (Ryder 1989) that is important as it provides a good historical account of developments in the human relationship with animals and of efforts to provide better protection for animals from human-inflicted suffering and *Animal Welfare: A Cool Eye Towards Eden* by John Webster published in 1995 (Webster 1995) that deals in a practical way with the recent and developing science of animal welfare that provides a rigorous basis for achieving improvements in the way animals are treated by humans.

The final three books selected deal with specific disciplines that in practice are of particular relevance to the actual achievement of reforms by either direct or legislative means to improve animal protection and reduce animal suffering. These books are *Ethics into Action: Henry Spira and the Animal Rights Movement* by Peter Singer published in 1998 (Singer 1998) in which the potential for direct action based on ethical considerations and practical activism is revealed and illustrated as well as some advice given to enable future campaigners to be more effective; *Political Animals: Animal Protection Politics in Britain and the United States* by Robert Garner published in 1998 (Garner 1998) that provides an analysis of political considerations that are so important in the achievement of legislative reforms to improve animal protection; and *Animal Welfare Law in Britain: Regulation and Responsibility* by Mike Radford published in 2001 (Radford 2001) that gives an authoritative detailed account and analysis of the role of law in the past, at the time of writing and in the future, particularly in Britain but with relevance for other jurisdictions, in the achievement of incremental improvements in the legal protection of animals.

It is seen that the books selected cover a range of important issues, background, and disciplines, and they provide a broad picture of many matters that are important for improving animal protection against human inflicted suffering. Moreover, the selected books are supplemented by a significant number of other related books that are mentioned or discussed. The above outline of the books selected for detailed discussion provides a road map of the way the book develops, and this is assisted by a separate chapter being devoted to each selected book as well as arrangement of the books in sequence of their date of publication. The study is not intended to provide traditional reviews of the books selected as these are available elsewhere, but in addition to giving an outline of the contents of the books selected, considerable emphasis is placed on their influence and potential.

The aim of the present work is to provide a broad understanding in a historical context of the diverse issues and disciplines involved in improving the welfare of animals whose quality of life is controlled or affected by the actions of humans. It is hoped the study will encourage further reading, particularly of the books selected or referred to in the book. More importantly, it would be good if the study influences, directly or indirectly, the actions or choices of individuals in ways that improve the lives of animals. And, of course, it would be particularly pleasing if the study

Table 1.1 Number of libraries shown on WorldCat as holding some edition of a selected book by country in April 2011

Year of first publication	Title of book	Author(s) of book	Number of libraries by country UK	USA	Australia	Total	Number of Google hits
1959	The Principles of Humane Experimental Technique	William Russell and Rex Burch	21	117	19	157	28700
1964	Animal Machines: The New Factory Farming Industry	Ruth Harrison	32	104	12	148	1.41×10^6
1975	Animal Liberation: Towards an End to Man's Inhumanity to Animals	Peter Singer	110	1,581	95	1786	333000
1989	Animal Revolution: Changing Attitudes Towards Speciesism	Richard Ryder	8	240	16	264	9090
1995	Animal Welfare: A Cool Eye Towards Eden	John Webster	10	86	35	131	61700
1998	Ethics into Action: Henry Spira and the Animal Rights Movement	Peter Singer	8	558	7	573	71200
1998	Political Animals: Animal Protection Politics in Britain and the United States	Robert Garner	11	153	5	169	27.5×10^6
2001	Animal Welfare Law in Britain: Regulation and Responsibility	Mike Radford	42	73	10	125	1.36×10^6

provides the catalyst for some people to become actively involved in one way or another in the animal movement and contribute in its quest to bring about much needed improvement in the treatment of animals by humans.

The question arises as to the current availability of the books selected. Some indication of this is provided by the number of libraries in different countries listed on WorldCat (2011) as holding the book. WorldCat is reputed to be the largest network of library-based content in the world. It embraces the catalogues of many major libraries around the world, including university and government libraries, but is not fully comprehensive in that it does not generally include, for instance, municipal libraries and school libraries. Nevertheless, it is a useful guide to the present availability of a particular book in many of the major libraries and of the number of such libraries in a specific country recorded as holding some edition of a particular book. In the case of the selected books, the number of libraries in the UK, USA, and Australia recorded on WorldCat as holding some edition of each of the books is shown in Table 1.1. A similar search can be made regarding the availability of a particular book in any other country. However, the number of libraries shown for a specific country as holding a particular book will depend on the number of libraries in that country that are included in the WorldCat database. A request to WorldCat for this information did not receive a response, so it is not possible to make a comparison between countries of the proportion of libraries in each country on the WorldCat database that hold a particular book. Such a comparison would be interesting.

The large number of libraries in each of the UK, USA, and Australia shown on WorldCat (Table 1.1) as holding some edition of *Animal Liberation* by Peter Singer provides some evidence that the book has been influential. Moreover, a reasonably large though smaller number of libraries are shown as currently holding *The Principles of Humane Experimental Technique* by Russell and Burch (1959) that proposed the now well-known *Three Rs*, and *Animal Machines* by Ruth Harrison (1964) that was the original catalyst for the movement against factory farming.

It is important to appreciate, however, that the number of libraries shown on WorldCat as holding a particular book cannot be regarded as an accurate reflection of its influence. This is because the number depends on many factors including the year of publication of a book, the price of the book, the extent to which it was promoted, and the number of editions published as well as on demand for the book. Moreover, the number of libraries currently holding a particular book might differ significantly from the number that held it at other times since its publication. Furthermore, there are many libraries not included on WorldCat, and many books are privately owned. Another factor is that the number of libraries holding a particular book relates only to its potential influence. Its actual influence will depend on whether it is read and if so on the impact of its contents. Moreover, the influence of a book is created not only by reading it but by the communication of its contents in other ways such as by being taught or written about. There does not appear to be an available record of the number of times a particular book has been cited, but even this could be misleading as to its influence. For instance, the well-known *Three Rs* and *Five Freedoms* are frequently referred to, but often, no indication is given as to their origins. The number of Google hits indicates that the world wide web is increasingly used to dissipate information about the books, since the two most recent books get the most hits, apart from Animal Machines. The search emphasizes the major legacy of Animal Machines, which was published nearly 50 years ago.

The books selected provide the ingredients for an integrated approach to improving animal welfare, and this will be explored in the concluding chapter. In any event there are many different means for improving the welfare of animals, and each of these should be used wherever possible to create a better life for animals. It is hoped the present study based on a selection of modern books provides an increased appreciation of how we have reached the stage we are at today in relation to animal welfare as well as of the complex, difficult, and often interrelated issues involved in providing improved protection for our fellow creatures.

Chapter 2
Experimentation on Animals

Abstract The number of animals used in laboratories for experimentation and routine testing of industrial products and medicines has grown rapidly in recent years. Fortunately the University Federation of Animal Welfare had the foresight to sponsor two scientists, Wiliam Russell and Rex Burch, to study methods of improving the management of laboratory animals to achieve a better welfare outcome. The results of their study were published in a book that quickly became recognised as *the* authoritative work on the topic. The book introduces the concept of the '3R's (Reduction in numbers, Replacement by alternatives, and Refinement of conditions for the animals). This approach was subsequently adopted by many laboratory animal scientists wordwide. The importance of this and other key texts in this field is discussed, alongside the scenario of growth in laboratory animal numbers and changing use patterns.

2.1 Introduction

Just over 60 years ago following extensive research, a book on the treatment of animals during experimentation was published at the end of the 1950s. It was

The Principles of Humane Experimental Technique by W.M.S. Russell and R.L. Burch published in 1959

2.2 Contents of the Book

In 1954 work began on a systematic research project (Russell and Burch 1959, p. xiii) sponsored by the Universities Federation for Animal Welfare (UFAW) on the progress of humane technique in the use of animals in the laboratory. The outcome of this research was published in 1959 in a now-celebrated book (Russell and Burch 1959) entitled *The Principles of Humane Experimental Technique* authored by William

E.N. Eadie, *Understanding Animal Welfare*, Animal Welfare 13,
DOI 10.1007/978-3-642-30577-1_2, © Springer-Verlag Berlin Heidelberg 2012

Russell DPhil(Oxon) (UFAW Research Fellow in the Department of Zoology and Comparative Anatomy at University College London) and Rex Burch (Honorary UFAW Research Assistant). The welfare of laboratory animals can be classified broadly into two categories (Russell and Burch 1959, p. 5), namely their husbandry when not under experiment and their treatment during the course of experiments. The former had been studied previously and the results published in the *UFAW Handbook on the Care and Management of Laboratory Animals*, while the latter aspect involving experimental technique itself formed the subject of the 1959 book. In that book a recurring theme is the close relationship between *humanity* and *efficiency* in experimentation.

The book *The Principles of Humane Experimental Technique* (Russell and Burch 1959) comprises two parts. Part 1 is entitled Scope for Humane Technique, and following an introduction, it deals with the concept of inhumanity, the ecology of experimental animals, and the sources, incidence, and removal of inhumanity. Russell and Burch (1959, p. 64) observe that the ways in which inhumanity can be and is being diminished or removed can be discussed under the three broad headings of *Replacement, Reduction*, and *Refinement.* The authors state (Russell and Burch 1959, p. 64):

> Replacement means the substitution for conscious living higher animals of insentient material. Reduction means reduction in the number of animals used to obtain information of given amount and precision. Refinement means any decrease in the incidence or severity of inhumane procedures applied to those animals which still have to be used.

Nevertheless, the authors observe there are clearly areas of overlap between these categories. Part 11 of the book entitled The Progress of Humane Technique comprises three major chapters covering Replacement (pp. 69–104), Reduction (pp. 105–133), and Refinement (pp. 134–153) that deal in detail with what have become to be known as the 3Rs. Finally, Russell and Burch (1959, pp. 154–168) discuss a number of factors they see as likely to govern progress in the achievement of humane experimental technique. These include sociological factors associated with humanity and efficiency, flexibility and communication, educational matters, and legal and semi-legal issues. Moreover, the authors observe that special organisations, namely The Laboratory Animals Bureau, The Animal Technicians' Association, and The Universities Federation for Animal Welfare (UFAW) have a role to play in communicating the message regarding the desirability of using humane experimental techniques. Of these organisations, the authors observe that UFAW was the only one at that time to have shown concern for the subject. In 1957, UFAW organised a symposium on Humane Technique in the Laboratory (Russell and Burch 1959, p. 167) that was held at Birkbeck College London and was chaired by Professor Peter Medawar CBE FRS (subsequently knighted and the winner of a Nobel Prize) who was Chairperson of the Scientific Advisory Committee of UFAW.

In the conclusion of their book on humane experimental technique, Russell and Burch (1959, p. 167) observed that the book provided only an outline of the subject and that it remained for others to elaborate on its contents. Finally, Russell and Burch (1959, pp. 167–168) expressed the hope that the book would stimulate some experimentalists to give special attention to the subject and for many others to be fully aware of its existence and possibilities. Moreover, they hoped the book would provide new workers in the field with a unified perspective of some of the most important aspects of the subject.

2.3 Influence of the Book

It is not known how extensively the book *Principles of Humane Experimental Technique* has been read, but the concept of replacement, reduction, and refinement for experimental animals expounded in the book is now well known and widely embraced, even though concerns have been expressed as to how well it is applied in practice. According to Russell (2005, p. 280) the book was hardly noticed for two decades until about 1980, but then the concept of the Three Rs suddenly took off and subsequent progress was astonishing and sustained.

In November 2005, a special issue of the journal *Animal Welfare* published by the Universities Federation for Animal Welfare (UFAW) was devoted to the Three Rs to commemorate the 50th anniversary of the first year of the research study by Dr William Russell and his assistant Mr Rex Burch into humane experimental technique that led to the publication in 1959 of their celebrated book (Russell and Burch 1959; Kirkwood 2005). The 2005 Special Issue of *Animal Welfare* contains an introductory paper by William Russell, then Professor of Sociology at the University of Reading, in which he reviews the past, present, and future of the Three Rs (Russell 2005, pp. 279–286). Significant developments in relation to the Three Rs described or alluded to in the review paper by Russell (2005, pp. 279–281) included:

1. Organisations specifically devoted to the Three Rs have been established in many countries. The first of these organisations was the Fund for the Replacement of Animals in Medical Experiments (FRAME) founded in Britain in 1969, and despite its name, it eventually became firmly committed to *all* Three Rs under the leadership of Professor Michael Balls. Indeed, this commitment was symbolised by FRAME in giving the name Russell and Burch House to its new building in Nottingham.
2. Other organisations devoted to the Three Rs included the Center for Alternatives to Animal Testing (CAAT) founded in 1981 in the USA, the Centre for Documentation and Evaluation of Alternative Methods to Animal Experiments (ZEBET) in Germany and the Japanese Society for Alternatives to Animal Experiments (JSAAE), each established in 1989, and the Netherlands Centre for Alternatives (NCA) founded in 1994.
3. The Three Rs have been embodied in the legislation of a number of countries as well as in the important 1986 European Union Council Directive 86/609/EEC.
4. Beginning in the Netherlands in 1985, courses on laboratory animal science and the Three Rs have been introduced to educate scientists intending to experiment on animals, and a Chair in Alternatives was established at the University of Utrecht. Subsequently, two such Chairs were created in Germany.
5. Following the creation of national platforms in Germany and The Netherlands in 1987 for coordinating ministries, universities, industry, and welfare societies in promoting the Three Rs, 11 such national platforms had been established by 2004, and these have proved a very useful type of organisation.
6. In 1990, the Humane Society of the United States (HSUS) established a new prize named the Russell and Burch Award for contributions to laboratory

animal welfare, and a number of distinguished scientists in the Three Rs field have received this international award.

7. A vitally important development for replacement was the establishment in 1991 of the European Centre for the Validation of Alternative Methods (ECVAM) at Ispra, Italy by the European Commission. In 1993 the centre began work with Professor Michael Balls as its Head, and under his direction and inspiration, great progress was made. ECVAM did much work on replacement itself, validated replacements of others, and achieved the necessary requirement of getting replacements accepted by regulators. Moreover, it developed a rational and rigorous form of validation. Indeed, this showed that in vitro methods are frequently better than in vivo ones.
8. In 1992, the book by Russell and Burch originally published in 1959 was reprinted.
9. A measure of worldwide activity, and a particularly exciting development for the Three Rs, has been the organisation of World Congresses involving hundreds of contributors from dozens of countries. The first of these was held in Baltimore in the USA in 1993, and this was followed by congresses in Utrecht in The Netherlands in 1996 and Bologna in Italy in 1999. The latter congress was hosted by Professor Michael Balls when he was Head of ECVAM. Further world congresses have been held since 1999. Moreover, the published proceedings of the congresses have made a major contribution to the literature in the field of the Three Rs.
10. The participants of the 1999 Bologna Congress agreed unanimously on the *Three Rs Declaration of Bologna*. Moreover, the major principle enunciated in the book by Russell and Burch (1959) of the close correlation between humaneness and scientific effectiveness was firmly asserted in the Declaration of Bologna. In 2000, the influential European Science Foundation (ESF) in its statement on *Use of Animals in Research* strongly endorsed the principles of the Three Rs with reference to the book by Russell and Burch (1959).
11. In 2004, Russell wrote (in English) the Preface to a book in Chinese on the Three Rs edited by He Zheng-Ming of the Beijing Society of Alternatives to Laboratory Animals, an organisation that was doing much to promote the Three Rs in China.
12. The use of computers and tissue cultures, especially in toxicity testing, have played an important role in replacement.

It is seen that substantial progress on many fronts has been achieved in relation to the Three Rs, a concept developed by Russell and Burch in their 1959 book, and indeed, the welfare of laboratory animals has been a major beneficiary. Nevertheless, there are potential opportunities for improvement in relation to use of the Three Rs, and some of these are outlined by Russell in his review paper (Russell 2005).

In contrast to *replacement* that opens up an enormous field of specific experiments and involves a huge amount of research, *reduction* requires only the proper application of well-established statistical principles (Russell 2005, p. 282). What is needed for effective reduction is appropriate education to ensure that these principles are applied correctly, and for this, a high level of experience in both statistics and biomedical science is required. In this regard, great reason for concern was found by the celebrated statistician in the field, Michael Festing, when he examined a large number of biomedical papers and discovered an enormous number of elementary errors in design and analysis that were sufficiently serious to invalidate experiments and so result in wasted animal use. In view of this problem, Festing has published numerous papers since 1994 in an attempt to get statistical methods applied properly (e.g. Festing 2009). Furthermore, it would be highly desirable for every experimenter not fully qualified in statistics to have an identified professional statistical advisor, and if made compulsory, this would result in reduced animal use.

On the matter of making animal populations more uniform, which influences the number of animals required, a long established principle discovered by ethologist Michael Chance in 1957 was that "some environments are more favourable to uniformity than others" and that the most uniform population results where animals are kept in "an environment optimal for their well-being" (Russell 2005, p. 283). In this regard, the goals of *reduction* and *refinement* correspond. Moreover, control of variation and approximation to uniformity can be achieved by manipulating either the genetic or the environmental situation and preferably both, as discussed by Russell (2005, p. 282). Another important aspect of *refinement* is the control of pain by the proper use of anaesthetics and analgesics, and possibly the greatest challenge for refinement entails the matter of humane end points. In toxicity testing as well as in testing the potency of biologicals, animals frequently end their life in severe distress as in the case of the notorious LD50 test (Russell 2005, pp. 283–284). Clearly, it is most desirable for physiological, biochemical, or immunological variables to be found that show the effects of a poison or infection before animal suffering begins and the animals then painlessly killed. In recent years much work has been done on the subject of humane end points, and the issue represents a highly important aspect of refinement.

In many countries researchers considering the use of animals in experimentation are required by law to examine opportunities for replacement, reduction, and refinement (the Three Rs) in order to minimise animal suffering. However, in practice, the procedures involved in doing this properly are extremely complicated and time consuming. In order to assist in this process, two papers were published in the Three Rs Special Issue of *Animal Welfare* to provide information on effective searching of the scientific literature for alternatives by using search grids to select appropriate databases (Hart et al. 2005, pp. 287–289) and the use of databases, information centres, and guidelines in planning research that might involve animals (Smith and Allen 2005, pp. 347–359). The latter paper includes an extensive appendix listing relevant information sources. These papers contribute to making the difficult task of identifying alternative procedures easier and more effective and in doing so result in improved animal welfare as well as compliance with the law. Moreover, there is scope for the use of professional information specialists to provide advice on the evaluation of alternatives relevant to a particular field of research.

An important development for the Three Rs, as well as for the influence of the 1959 book by Russell and Burch that advocated them, was the publication in 2009 by Professor Michael Balls of the Fund for the Replacement of Animals in Medical

Experiments (FRAME) of *The Three Rs and the Humanity Criterion: An Abridged Version of the Principles of Humane Experimental Technique by WMS Russell and RL Burch (1959)* (Balls 2009). This new book is an abridged, annotated, and updated version of the 1959 book by Russell and Burch edited by Michael Balls (Boyd 2010, p. 197). In the book, Balls has abridged some currently less relevant parts of the original book, and he explains why and how other parts of the book have been superseded by later developments. Moreover, Balls has provided explanatory notes in relation to literacy and phraseology to make the work more readable in today's world, particularly for those whose first language is not English. The new work enables a clear understanding of the clarity with which Russell and Burch defined and illustrated each of their three principles as well as the interrelationship between them and of the philosophical and scientific analysis of other relevant concepts explored by the original authors (Boyd 2010, p. 197). Moreover, the title of the abridged version emphasises the *humanity criterion* that was so important to Russell and Burch. The abridged book by Balls is especially important as it will make available to a wide and international audience the seminal work on the Three Rs published by Russell and Burch 50 years ago, and it should contribute to greater understanding and practice of the principles expounded by the original authors.

Michael Balls has contributed enormously over many years to developments related to the Three Rs, particularly replacement, through his work at the European Centre for the Validation of Alternative Methods (ECVAM) and at the Fund for the Replacement of Animals in Medical Experiments (FRAME) that published Balls (2009), and he was eminently suited to produce the abridged version of Russell and Burch (1959). As a result of publication of the new book, the influence of the classic work of Russell and Burch will continue to grow in the future, and in doing so it will continue to contribute significantly to a more humane world for experimental animals. Ironically, it was not until the late 1980s that Balls discovered the 1959 book of Russell and Burch, even though he had been deeply committed to replacement for about 15 years (Boyd 2010, p. 197), and it is likely that many other people, who are aware of the Three Rs, would have found out about them other than by reading the seminal work in which they were expounded by Russell and Burch.

The Three Rs have been recognised widely by governments around the world, and they have been incorporated into national, regional, and international legislation, regulation, or policy. For instance, in Australia, the *Australian code of practice for the care and use of animals for scientific purposes* (Commonwealth of Australia 2003) specifies that the Three Rs be given consideration in the conduct of scientific research and teaching activities involving animals. This code is incorporated in legislation of the individual states and territories of Australia as in the case of the Queensland *Animal Care and Protection Act* 2001 (Reprint No 2A 2003) under which compliance with the code is made mandatory in Chap. 4 of that act entitled: Using animals for scientific purposes. Regionally, in the European Union, the Three Rs underlie important provisions in EC legislation designed to protect laboratory animals (Eurogroup for Animal Welfare 2004, pp. 91–93), and under Directive 86/609/EEC Member States are obliged to encourage research into alternatives. Moreover, under this 1986 Directive, the European Community was committed to establish the European Centre for the Validation of Alternative Methods (ECVAM) that began its work in 1993 (Hughes 1995, pp. 14, 15, 32), and since then, ECVAM has played an important and internationally recognised role in relation to alternatives (Eurogroup for Animal Welfare 2004, pp. 91–93). At a global level the World Organisation for Animal Health (OIE) has been developing animal welfare standards (Department of Agriculture, Fisheries, and Forestry 2005, pp. 33–34) that follow agreed *OIE Guiding Principles for Animal Welfare*. Among these guiding principles is acceptance that:

> The internationally recognised Three Rs provide valuable guidance for the use of animals in science.

It is hoped that the work of OIE in establishing animal welfare standards embodying the Three Rs will contribute to greater appreciation of animal welfare issues and result in improved animal welfare outcomes for laboratory animals internationally.

Even though there is recognition of the Three Rs at various levels of government, reservations have been expressed as to how well this recognition is translated into practice. For instance, the Stocktake of the Australian Animal Welfare Strategy Education and Training Working Group (2008, pp. 2–4) identified a number of key insights for the various animal sectors and observed in relation to animals used in research and teaching that welfare could be improved through better training of care staff, greater enforcement of existing legislation, and more widespread teaching of animal welfare and ethics. Regionally, in the European Union, the work of the European Centre for the Validation of Alternative Methods (ECVAM) was praised by Eurogroup for Animal Welfare (2004, pp. 91–93), but Eurogroup regarded the political and financial support for ECVAM as inadequate and advocated that more funding should be allocated to support its work. Moreover, Eurogroup saw it as essential that the Commission and Member States act expeditiously to implement the adoption of alternative testing strategies developed by ECVAM.

The question needs to be asked as to how well *in practice* the relevant legislative provisions relating to the Three Rs are being effectively applied and how seriously alternatives to the use of animals are properly considered. Indeed, according to Malcolm Caulfield (2008, pp. 158–159) many laboratory animals are made to suffer by biological scientists motivated more by job security and prospects than good science, and this suggests the need for greater accountability. Research conducted by Siobhan O'Sullivan (2006, pp. 1–15) on transparency and research regulation in Australia suggested the absence of an adequate system of transparency, and she concluded that any enhanced transparency was likely to come only as a result of change in public policy. Greater transparency should result in a better informed public, and this could help increase the effectiveness of government regulation designed to reduce the suffering of experimental animals. A recent opinion article by Simon Festing (2010, pp. 22–23) published in the *New Scientist* expressed the view that there were too many poorly designed, conducted, reported, and reviewed animal studies and that biomedical research needs to improve its act so that animals are not made to suffer needlessly in valueless experiments. Moreover, this is the view of a person who supports the carefully planned and well-regulated use of animals for scientific and medical advancement while recognising that such research is controversial.

These observations indicate the need for improved education in animal welfare and experimental design, more effective enforcement of regulations, and greater personal commitment on the part of researchers to properly consider the suffering of animals. Indeed, a similar personal commitment is considered by Professor Jeffrey Sachs as an important factor in the future alleviation of world poverty (Sachs 2005, pp. 364–368). There appears to be considerable scope for the development of a more enlightened and compassionate attitude among many researchers in relation to the welfare of the animals for which they are responsible, and if achieved, this could play an important part in reducing the suffering of experimental animals and improving proper application of the Three Rs.

It is appropriate to consider the scale of animal experimentation in terms of the number of animals involved while remembering that the welfare of each individual animal is important. Statistics of scientific procedures on living animals in Great Britain published by the Home Office for 1988–2006 are illustrated and discussed in an article in *FRAME News* (Trigwell 2007, pp. 10–11). The statistics show that total annual procedures on animals decreased significantly for about a decade from 1988 then rose again gradually to reach a 15-year high in 2006 when 3,012,032 procedures were carried out in Great Britain. A large proportion of these procedures involved the use of rats and other rodents. During the period from 1995 to 2006 the proportion of animals used for fundamental biological research remained steady at around 30 %, whereas there was a significant decrease in the percentage of animals used for applied research and a significant increase in the percentage of animals used for breeding. The article concluded (Trigwell 2007, p. 11):

> The continued rise in the levels of animal experimentation is an ongoing concern for FRAME. There is an urgent need for a review of why certain types of experiments are permitted, which takes into account the availability of alternatives and the scope for using existing information to enrich our understanding of human and animal physiology and to improve animal welfare. There needs to be greater dialogue between all interested parties, including industry, academia, and the Government, and better application of the Three Rs by all scientists, if this worrying trend is to be prevented from escalating out of control.

On a global scale, it is reported by Clive Phillips (2009, pp. 156–157) that the annual number of vertebrate animals used in experimentation could be as high as 75–100 million with the USA, the European Union, and Japan being large users of laboratory animals. It is acknowledged by Phillips (2009, p. 156) that for various reasons it is difficult to make reliable estimates of the number of animals used. Nevertheless, it is clear that the number of animals involved in experimentation around the world each year is huge, and this results in an enormous amount of animal suffering.

Mention should be made that leading animal rights exponent and philosopher Tom Regan (1983, 1985) argues that it is fundamentally wrong for humans to view animals as resources for their use. It is asserted by Regan (1985, pp. 23–24) that the granting of rights to animals rests on their *inherent value* that belongs equally to all those who are *subjects to a life*. Among the goals of the animal rights movement (Regan 1985, p. 13) is total abolition of the use of animals in science. Recently, J.M. Coetzee, winner of the Nobel Prize in Literature, wrote the foreword to the book *Second Nature: The Inner Lives of Animals* written by Jonathan Balcombe (2010). Coetzee acknowledges that Balcombe is "critical of the treatment routinely meted out to animals in research laboratories" but comments that he does not ask the truly radical antivivisection question: Can the suffering and death of animals in laboratories ever be justified, even if it leads to advances in scientific knowledge? (Coetzee 2010, p. xii). For almost three decades, Regan has argued against the morality of animal experimentation, and Coetzee questions whether it can ever be justified. However, it is clear that animal experimentation continues to take place on a large scale, and this situation is unlikely to change dramatically in the foreseeable future, if at all. Thus, it is as important as ever that the Three Rs proposed by Russell and Burch 50 years ago be properly understood and effectively applied so that animal suffering in experimentation is kept to a minimum. Even though a great deal of progress has been made over the years in relation to the Three Rs, there is considerable reason for disquiet and great scope for their proper application. This presents an enormous challenge for the future.

2.4 Potential for the Book

There is substantial opportunity in the future for the principles of replacement, reduction, and refinement (Three Rs) proposed and expounded by Russell and Burch (1959) to achieve more humane experimental technique in the use of animals in the laboratory. This will be assisted greatly by publication of the abridged version of Russell and Burch (1959) by Michael Balls (2009) that will make the original ideas of Russell and Burch more widely accessible and readable, especially to those whose first language is not English.

Finding a suitable *replacement* for the use of animals in research and testing can present a significant challenge, and this can be made easier by using available published material on effective searching of the literature to find suitable alternatives to live animals as well as the use of databases, information centres, and guidelines for planning research. Moreover, it would be highly desirable, if not mandatory, for researchers proposing to use animals in their work to consult appropriate professional information specialists regarding possible alternatives to animal use. In the case of *reduction*, which depends on the proper use of statistics, it should be a requirement for researchers using animals to be given at least an elementary grounding in statistics so they understand basic statistical principles. Even though this is often done, there is reason for great concern in that a significant proportion of research studies involving animal use appear to be badly planned or involve flawed statistical analysis (or both). Consequently, there is a strong case for making it mandatory for a properly qualified statistician to be formally connected to each research study that involves animal use to ensure proper experimental design and the correct analysis of results. This would greatly reduce the potential of animal suffering for no avail. On the matter of *refinement*, researchers using animals need to be fully aware of the issue of pain. Moreover, in certain areas of animal use, as in the testing of cosmetics or the potency of biologicals, there is great need to establish suitable humane end points so as to circumvent severe animal suffering. Reduction and refinement in animal use can result also from greater uniformity in the animals used in an experiment, and this can be achieved by means of environmental and/or genetic manipulation. Thus, there will be considerable scope in the future for proper application of the Three Rs to reduce the suffering of animals in experimental situations.

A variety of other measures can contribute also to more effective use of the Three Rs. These include:

1. Wider education of biological and medical scientists in animal welfare and ethics generally so they recognise and think of animals as *sentient beings* and more specifically in the Three Rs, statistical principles, and the legal obligations of experimenters using animals.
2. General adoption in all relevant jurisdictions of tight legal regulations to protect experimental animals and more effective enforcement of these regulations. There is scope for such regulations at national, regional, and international levels, and there needs to be much more than just "lip service" given to the Three Rs.
3. Creation of greater transparency[1] as a matter of public policy so the community is more aware of the issues and of what is happening. Such transparency could contribute to exposing abuse where it occurs.
4. Greater recognition that the finding of alternatives to animal use is an important area of research and for such research to be regarded as a prestigious field for investigation. This can be assisted by the creation of awards such as the Russell and Burch Award established by the Humane Society of the United States (HSUS) in 1990 (Russell 2005, p. 280) and the award initiated in 2003 in New Zealand by the National Animal Ethics Advisory Committee (NAEAC) to recognise the implementation in that country of the Three Rs (National Animal Ethics Advisory Committee 2004, p. 18).
5. Dissemination of published papers produced by specialised organisations such as FRAME, the proceedings of global conferences on the Three Rs, and the research literature generally, each of which provides a wealth of relevant information. However, there needs to be an effective means of cataloguing or classifying this material so it is made more readily accessible to researchers, and in this information, specialists could help. Moreover, there could be more widespread publication of negative results to avoid repetition of research and animal wastage.
6. Effective cooperation between academia, industry, and government in an attempt to avoid wasteful duplication of research using animals and to ensure that any discovered alternatives to animal use are expeditiously validated so they are accepted by regulators. Such cooperation needs to be on an international basis.
7. Development of a culture of *personal commitment* among experimental scientists and others to avoid animal use wherever possible and in any event to keep animal suffering to a minimum. This could be engendered by proper recognition of animals as sentient beings and the development of enlightened attitudes of compassion and respect for them.

[1] In industrial countries, many animals are used for regulatory testing of products by the chemical and pharmaceutical industries, which the public is not aware of. Transparency is needed in the uses of animals, as well as how they are used.

Thus, with appropriate commitment and application, there are various opportunities in the future that can effectively influence application of the Three Rs proposed by Russell and Burch (1959), and in this, the abridged version of their book by Balls (2009) is likely to make a valuable contribution. The number of animals used in experimentation remains large, and a concerted effort needs to be made on many different fronts by all those involved to ensure that the Three Rs are properly applied so the extent of animal suffering in experimentation is substantially reduced. For this to be achieved, strong personal commitments need to be made by those involved, whether in university, industry, or government.

2.5 Some Other Books

According to Russell (2005, p. 280), Andrew Rowan is the leading historian of laboratory animal welfare, and his book *Of Mice, Models and Men* was published in 1984 (Rowan 1984). In 1975 Richard Ryder published *Victims of Science: The Use of Animals in Research* with the aim of providing documentary evidence of humans' mistreatment of animals for research purposes and to suggest reforms. Ryder makes reference in his book (Ryder 1975, p. 238) to the book by Russell and Burch published in 1959 and observes that the book was publicised by the RSPCA in Britain. Although concerned with humane education rather than animal experimentation, another book *From Guinea Pig to Computer Mouse: Alternative Methods for a Progressive, Humane Education* by Jukes and Chiuia (2003) is a valuable text on teaching alternatives to animal use in education. The book presents a comprehensive review of available literature and resources related to alternatives to animals for teaching purposes (Hart 2004, p. 257), and this book could well be a valuable catalyst for instilling more humane attitudes among future experimental scientists. In 2004 *The Welfare of Laboratory Animals* edited by Eila Kaliste (2004) was published as a book in what has become the Springer Animal Welfare Series. This book contains a chapter on Experimental Procedures written by David Morton (2004, pp. 81–115) that deals with general principles and recommendations and covers handling, dosing, sampling, anaesthesia, analgesia, euthanasia, and humane end points in research. In addition, the book contains chapters on the welfare of specific species of experimental animals including mice, rats, guinea pigs, and rabbits. Such books provide useful further reading related to humane experimental technique and application of the Three Rs expounded by Russell and Burch in 1959.

2.6 Overview

The animal rights movement, of which philosopher Tom Regan (1983) and lawyer Gary Francione (1995) are leading exponents, is opposed entirely to the use of animals in experimentation and testing. The reality is that such activities can be expected to continue in the future, for some time at least, as evidenced by their current prevalence. This is in spite of campaigning over many years by both the antivivisectionist movement and animal rights activists to stop them. An attempt to bring greater "humanity" into the actual conduct of animal experimentation was made by Russell and Burch (1959) in their celebrated book that proposed the Three Rs of replacement, reduction, and refinement. The Three Rs are now well known, and much has been done to implement them as described earlier. However, in order for the Three Rs to be fully effective:

1. All those involved in animal experimentation need to be well educated about the meaning of the Three Rs, have a compassionate attitude towards animals, and recognise the need to minimise their suffering.
2. Ensure the conduct of any experimentation involving animals is really justified, establish that similar research has not been done previously, make a genuine attempt to explore the use of alternatives, and if necessary consult information specialists to help search for these.
3. Apply proper statistical techniques both in experimental design to ensure the minimum (but not too few) animals are used to provide meaningful results and in the analysis of results so they are properly interpreted to avoid animals being wasted needlessly.
4. Make sure that appropriate anaesthetics or analgesics are used to reduce the suffering of any animals used in experimentation.

This is the ideal if animal experimentation is to continue.

However, there is much evidence as discussed that the reality often falls short of this ideal, as when unjustified or flawed experiments are conducted. Moreover, the regulation of animal experimentation is not always enforced even though the Three Rs are enshrined in legislation. In order to make the Three Rs more effective, there is need for more widespread animal welfare education (preferably with course evaluation) of animal experimenters and regulators so as to engender a proper understanding of and compassion towards animal sentience and suffering, greater transparency as a matter of public policy so that unjustified or unscrupulous animal experimentation is exposed, and more effective enforcement of legislation designed to regulate animal experimentation that will both deter and detect unlawful practices. Furthermore, animal experimentation is an international issue, and the World Organisation for Animal Health (OIE) that is formally guided by the Three Rs has an important role to play in relation to international standards. Much good work of various kinds has been done over the years by many organisations and individuals in relation to the application and regulation of the Three Rs proposed by Russell and Burch in 1959, but this should not be used as a camouflage to engender a false sense of security that all is well in the world of animal experimentation when it is not. Indeed, there is much that can be done to improve it, and this represents a fruitful and worthwhile area for future research.

Chapter 3
Intensive Farming Industry

Abstract Intensive animal farming has expanded rapidly in the last fifty years in response to an expanding population that naturally preferred cheap animal products. However, the farmers and politicians underestimated the extent of public concerns and one of the first books to alert them to this was Animal Machines by Ruth Harrison, published in 1964. The impact of this and other books is discussed, as well as the progress in reversing the intensification of animal production systems.

3.1 Introduction

Following the Second World War, there was a rapid growth in industrial animal agriculture known as factory farming, and the animal suffering involved in these new agricultural practices was vividly exposed by Ruth Harrison in 1964 with the publication of her book

Animal Machines: The New Factory Farming Industry by Ruth Harrison published in 1964

3.2 Contents of the Book

By publishing her now famous book *Animal Machines*, Ruth Harrison intended to inform the public and stir passion against the disturbing practices used in the then newly established factory farming industry, and her book contains substantial authoritative scientific, economic, and legal evidence to support her arguments. In the Foreword to *Animal Machines* Rachel Carson, author of *Silent Spring*, wrote (Carson 1964, p. vii):

> The modern world worships the gods of speed and quantity, and of the quick and easy profit, and out of this idolatry monstrous evils have arisen. Yet the evils go long unrecognised. Even those who create them manage by some devious rationalising to blind themselves to the harm they have done society. As for the general public, the vast

E.N. Eadie, *Understanding Animal Welfare*, Animal Welfare 13,
DOI 10.1007/978-3-642-30577-1_3, © Springer-Verlag Berlin Heidelberg 2012

> majority rest secure in a childlike faith that 'someone' is looking after things – a faith unbroken until some public-spirited person, with patient scholarship and steadfast courage, presents facts that can no longer be ignored.
>
> This is what Ruth Harrison has done. Her theme affects practically every citizen, for it deals with the new methods of rearing animals destined to become human food. It is a story that ought to shock the complacency out of any reader.

Rachel Carson concluded (Carson 1964, p. viii):

> Although Ruth Harrison's book describes in detail only the conditions prevailing in Great Britain, it deserves to be widely read also in those European countries where these methods are practised, and in the United States where some of them arose. Wherever it is read it will certainly provoke feelings of dismay, revulsion, and outrage. I hope it will spark a consumers' revolt of such proportions that this vast new agricultural industry will be forced to mend its ways.

In his Preface to *Animal Machines*, Sydney Jennings MRCVS, Past-President of the British Veterinary Association, wrote in relation to the moral issue of the limit to where intensive farming can go (Jennings 1964, pp. ix–x):

> How far will man claim the right to inflict bodily discomfort and cause mental anguish to birds and animals if factory farming methods appear to require it? There are some who claim that we have reached the stage when intensive methods can go no further; there are many who believe they have already gone too far. Who is to be the judge? Ruth Harrison has wisely refrained from frequently expressing her own opinions. The fact that she has written *Animal Machines* leaves the reader in no doubt as to where she stands, but the author has let agriculture tell its own story in that hundreds of references from various forms of agricultural literature have been quoted. The reader can be his or her own judge.

Jennings observed also the relevance that "after the Second World War millions of German people asserted they knew nothing of what transpired in concentration camps" and that Ruth Harrison "is doing her utmost to prevent ignorance in regard to animal life in Britain" (Jennings 1964, p. xi). Moreover, Jennings commented that Harrison in her book is concerned not only with the issue of cruelty in factory farming but draws attention also to the potential danger to humans from the absorption of hormones, antibiotics, and other substances through eating meat. On the matter of cruelty, Jennings (1964, p. x) warned against the use of false premises to indicate the absence of animal discomfort, pain, or mental anguish and illustrates this by reference to the eating and putting on weight by veal calves kept in solitary confinement and darkness.

In order to help explain the meaning of factory farming, Ruth Harrison in her introduction of *Animal Machines* (Chap. 1, pp. 1–8) quotes the following words of an expert, Dr Preston of the Rowett Research Institute, published in *Farmer and Stockbreeder* on 19 December 1961 (Harrison 1964, p. 1):

> Rapid turnover, high-density stocking, a high degree of mechanisation, a low labour requirement, and efficient conversion of food into saleable products, were the five essentials for a system of animal production to be called intensive.

Such conditions are in stark contrast to the natural surroundings in which the public are led to believe food animals are reared. Such lack of transparency and the cruelty involved in rearing food animals in intensive farming situations, coupled with the assertion that what is being produced is what the public wants, motivated

Ruth Harrison to write *Animal Machines*. Subsequently, as the research for her book evolved, she discovered that the food produced in factory farming operations was both inferior and dangerous. In *Animal Machines,* Ruth Harrison (1964) considers broiler chickens (Chap. 2, pp. 9–26), poultry packing stations (Chap. 3, pp. 27–36), battery birds (Chap. 4, pp. 37–61), veal calves (Chap. 5, pp. 62–88), and other intensive units covering broiler beef, rabbits, and pigs (Chap. 6, pp. 89–97). Following these chapters on specific animal types and practices involved in intensively produced food, Harrison provides a powerful pictorial summary of the new factory farming, as well as (in contrast) of some traditional farming environments, in 24 photographic illustrations (Figs. 1–24) that are accompanied by appropriate explanations. These photographs reinforce in an appallingly graphic way the conditions of animals kept in the intensive farming situations described in the previous chapters. Subsequently, Harrison discusses the basis of quality (Chap. 7, pp. 99–109), quantity versus quality (Chap. 8, pp. 110–141), and finally cruelty and legislation (Chap. 9, pp. 142–167).

In the conclusions (Chap. 10, pp. 168–179), Ruth Harrison questions the economic arguments used in favour of factory farming and examines the quality and humanitarian arguments against it. She makes recommendations to protect both humans and animals from the undesirable practices described in *Animal Machines*. In the case of protecting human health, Harrison (1964, p. 176) advocates safeguarding the food we eat from additives not thoroughly tested and understood, the provision of information necessary to allow consumers the option of choice in relation to the food they eat, and a reassessment of true quality in food. On the matter of humanitarianism, Harrison recommends a new charter for animal welfare that includes (Harrison 1964, p. 178):

1. The complete abolition of battery cages for laying hens.
2. The complete abolition of the intensive methods now used in veal production. Both these practices are as unnecessary as they are undesirable.
   The abolition of egg and veal production by these methods in this country would have to be reinforced by the abolition of *imports* of eggs and veal produced by corresponding methods. It would be a poor gesture to ban the methods in our own country whilst supporting by imports the same methods used elsewhere.
3. I would like to see specific legislation banning the rearing of animals on deficiency diets. This would preclude food designed to produce anaemia such as is fed to veal calves and the possibility of blindness in barley beef calves.
4. Permanent tethering should be banned.
5. Slats should be banned.
6. The keeping of animals in dim light or darkness should be banned. This is a sign of bad husbandry and is completely unnecessary where conditions are satisfactory for the animals.

Harrison recognises that legislation alone will not provide animals with adequate protection and that there needs to be reassessment of our basic attitude towards animals bred solely to benefit humans (Harrison 1964, p. 178). Moreover, in relation to this, she stipulates the need for education throughout the entire fabric of our society.

3.3 Influence of the Book

The immediate impact of the publication of *Animal Machines* was front page coverage in every major newspaper in the UK, and extracts of the book were serialised in *The Observer* (Singer 1975, p. 137; de Fraga 2000, p. 26). This resulted in widespread public reaction as well as refusal by the public to accept assurances from the British Minister of Agriculture that all was well. Consequently, 6 weeks after publication of the book, the Minister of Agriculture appointed an expert committee to investigate intensive farming practices. Ruth Harrison was made a member of the committee that was chaired by eminent zoologist Professor F.W. Rogers Brambell and comprised other experts in veterinary science, animal husbandry, and agriculture among whom was W.H. Thorpe, Director of the Department of Animal Behaviour at Cambridge. The committee received evidence from a wide range of sources and visited a large number of livestock establishments throughout Britain as well as in Denmark and Holland for comparative purposes (Singer 1975, p. 138). The findings and recommendations of the investigation were published in December 1965 in an 85-page official report known as the Brambell Report (Brambell 1965).

The Brambell Committee enunciated the following fundamental principle for governing the degree to which an animal may be confined (Singer 1975, p. 139):

> In principle we disapprove of a degree of confinement of an animal which necessarily frustrates most of the major activities which make up its natural behaviour .... An animal should at least have sufficient freedom of movement to be able without difficulty to turn around, groom itself, get up, lie down and stretch its limbs.

The minimal standards that an animal should be able to turn around, groom itself, get up, lie down, and stretch its limbs became known as the "Five Freedoms". The Brambell Report commented on each of the major intensive farming activities in Britain, namely laying hens, table chickens (broilers), pigs, and veal calves as outlined by Singer (1975, pp. 139–145), and Singer also compared the situation in Britain as described in the report with that in the USA, where conditions for intensively farmed animals were considerably worse. Moreover, the Brambell Report stressed the importance of good stockmanship when intensive farming systems were used.

The findings of the Brambell Committee confirmed the conditions of intensively farmed animals in Britain exposed by Ruth Harrison in *Animal Machines*, although the recommendations of the committee were a compromise between the conditions the committee regarded as necessary to properly safeguard farm animal welfare and standards the committee thought the government and farmers might be prepared to accept (Singer 1975, p. 166). Nevertheless, the government refused to implement the recommendations of the committee. Indeed, the setting up of the Brambell Committee might have been a tactic to postpone action until strong public feeling against intensive farming subsided (Singer 1975, p. 137). However, the work of the Brambell Committee, of which Ruth Harrison was a member, did lead to the enactment in Britain in 1968 of the *Agriculture (Miscellaneous Provisions) Act* 1968 that covered general farm animal welfare (de Fraga 2000, p. 26). Moreover, the

Brambell Report that arose from the publication of *Animal Machines* provided a valuable document on which future progress could be built to improve the conditions of intensively farmed animals. This was particularly evident in Europe, although the report had relevance to intensive farming all over the world. In addition, Ruth Harrison's book *Animal Machines* was translated into seven languages, and this extended the scope for its influence.

Ruth Harrison encouraged research into intensive farming practices and maintained an active involvement in animal protection organisations that provided a vehicle for the propagation of knowledge and ideas she gained in researching and writing *Animal Machines*. In 1967, she was instrumental in the formation of the Farm Animal Care Trust that funded research "to make animals' lives better", and this facilitated important research over a 6-year period by Professor John Webster of Bristol University into the housing of veal calves (de Fraga 2000, p. 27). Indeed, subsequently, it became illegal in the UK to use small individual crates for veal calves. Moreover, the Council of Europe in 1976 adopted the *Convention for the Protection of Animals kept for Farming Purposes* that outlined general principles covering the basic welfare of farm animals (de Fraga 2000, pp. 26–27), and for many years, Ruth Harrison served as a member of the Standing Committee for that convention, initially as representative of the World Society for the Protection of Animals (WSPA) and subsequently of Eurogroup for Animal Welfare (Eurogroup). Moreover, she was involved in the political process by attempting to obtain a better life for farm animals through her long-standing membership of the United Kingdom Farm Animal Welfare Advisory Committee and its successor the Farm Animal Welfare Council. Ruth Harrison always adopted an ethical stance and advocated a precautionary approach that gave animals the benefit of doubt.

For many years the minimal "Five Freedoms" enunciated in the Brambell Report dominated discussion of animal welfare issues in Europe (Webster 1995, p. 11). However, John Webster, Professor of Animal Husbandry in the Department of Veterinary Science at the University of Bristol and a member of the UK Farm Animal Welfare Council (FAWC)[1], regarded the traditional "Five Freedoms" as being too obsessed with space requirements and concentration on only one aspect of animal behaviour, namely comfort seeking. They did not include other aspects of animal welfare such as those relating to food, health, and security. Accordingly, Webster advocated a more comprehensive analysis of *all* the factors influencing farm animal welfare, while preserving the concept of the "Five Freedoms", and this suggestion evolved over time into what are now commonly known as the *Five Freedoms* for animals that are widely accepted. For instance, RSPCA Australia in its policy document specifies the following Five Freedoms for animals (RSPCA Australia 2008, p. 6):

[1] The Brambell Report led to the establishment of the Farm Animal Welfare Advisory Committee to British government in 1967, which became the Farm Animal Welfare Council in 1979.

1. Freedom from hunger and thirst:
   By ready access to fresh water and a diet to maintain full health and vigour
2. Freedom from discomfort:
   By providing an appropriate environment including shelter and a comfortable resting area
3. Freedom from pain, injury, or disease:
   By prevention through rapid diagnosis and treatment
4. Freedom to express normal behaviour:
   By providing sufficient space, proper facilities, and company of the animal's own kind
5. Freedom from fear and distress:
   By ensuring conditions and treatment which avoid mental suffering

These are almost identical to the revised *Five Freedoms* specified in 1993 by the UK Farm Animal Welfare Council (Webster 1995, p. 11). According to Webster, these Five Freedoms can be used to evaluate, in a systematic and comprehensive way, the welfare of animals in a range of different environments such as to compare the welfare of laying hens kept in battery cages with those on free range. Moreover, the Five Freedoms provide a useful yardstick against which legal provisions designed to protect animals can be measured (Eadie 2009, pp. 4–5).

Ruth Harrison was actively involved with Eurogroup for Animal Welfare, and its campaigns over the years have contributed to new legislation as well as to changes in existing legislation in the European Union (EU) to improve animal welfare and reduce animal suffering (de Fraga 2000, p. 27; Eurogroup for Animal Welfare 2003, p. 2; Eurogroup for Animals 2008, pp. 9, 10, 12). Among the achievements that followed from Eurogroup campaigns were the 1997 calves Directive imposing a ban on individual calf crates, the 1998 Directive on general farm animal welfare, the 1999 laying hens Directive imposing a ban by 2012 on traditional battery cages, and in 2001 (after the death of Ruth Harrison in 2000) a requirement for the compulsory labelling of eggs according to the method of production and the 2001 Directive banning the use of individual sow stalls. The tethering of pigs in the EU ceased in 2006, and the use of narrow sow stalls throughout pregnancy will be abolished by 2013. In 1987 Harrison gave the annual Hume memorial lecture entitled *Farm Animal Welfare: What, If Any Progress?*, and during her lifetime, she expressed frustration at the slowness of reform that she attributed largely to government inaction and conflicting policies (de Fraga 2000, p. 27). Indeed, the ban in the European Union on the use of traditional battery cages for laying hens did not come into effect until 2012, almost 50 years after the publication of *Animal Machines* in 1964, and the EU is ahead of most other jurisdictions in this regard. Moreover, after 2012, "enriched" cages that have a perch, nests, and scratching areas are permitted for laying hens in the European Union, although in 2001 the German Parliament went a step further and made a unilateral decision to allow only free-range and barn systems for laying hens after 2012 (de Fraga 2002, p. 25).

It was observed by Ruth Harrison in her recommendations in *Animal Machines* (Harrison 1964, p. 178) that an abolition of battery cages for laying hens would have to be reinforced by the abolition of the *import* of eggs produced by the same methods. A problem she could not have envisaged was the effect of the narrow rules of the World Trade Organisation (WTO) that so far have not acknowledged animal welfare as a factor in determining what are "similar" products in terms of its process and production methods (PPMs) (RSPCA United Kingdom and Eurogroup for Animal Welfare 2001; Stevenson 2002). To the WTO an egg is an egg whether from battery-caged or free-range hens, and animal welfare is not a consideration. Indeed, there is need for the WTO to properly embrace animal welfare considerations (as well as, incidentally, social justice and environmental protection) into the international trading regime, and at present, the WTO rules provide a potential threat to the ban on battery cages in the European Union from 2012.

In recent years the World Organisation for Animal Health (OIE) has become involved in the development of scientifically based animal welfare standards (Department of Agriculture, Fisheries and Forestry 2005, pp. 33–34), and this could provide the vehicle for achieving improved farm animal welfare standards globally. This would reduce the effect of import competition from animal products produced under less favourable animal welfare conditions, although this is likely to take considerable time. Nevertheless, it is pleasing that the *OIE Guiding Principles for Animal Welfare*, with which any developed animal welfare standards have to be consistent, include (Department of Agriculture, Fisheries and Forestry 2005, pp. 33–34):

> The internationally recognised "Five Freedoms" provide valuable guidance in animal welfare.

These "Five Freedoms" developed from the original Five Freedoms enunciated by the Brambell Committee that was set up following the publication of *Animal Machines*.

It was recognised by Ruth Harrison that legislation alone would not provide all the answers to farm animal protection and that there needed to be a reassessment of basic human attitudes assisted by knowledge of how our food is produced. Indeed, this has been happening as a result of greater publicity, and it is evidenced, for instance, by a dramatic drop in veal consumption in the USA over the last 30 years or so as consumers became aware that calves were separated from their mothers soon after birth, deliberately kept anaemic, deprived of roughage and exercise, and housed in stalls too narrow for them to turn around (Singer and Mason 2006, p. 2). Moreover, McDonalds in the UK, in response to consumer preference for humanely produced products, more than 5 years ago changed egg suppliers to those dealing only in free-range eggs (Sherman et al. 2005, p. 7), and more recently, in Australia, in response to an RSPCA Australia *choose wisely* campaign, the Hilton Hotel in Adelaide converted to using only free-range eggs (RSPCA South Australia 2009, p. 6; Innes 2009, p. 15). Indeed, education and increasing public awareness that were regarded as so important by Ruth Harrison are playing a significant role in bringing about improvements in farm animal welfare and reducing the impact of intensive farming practices in creating animal suffering. However, much remains to be done.

3.4 Potential for the Book

In Europe, considerable scope remains for further improvement in the welfare of intensively farmed animals. Moreover, in other countries such as the USA and Australia as well as in developing countries, the welfare conditions for intensively farmed animals are less favourable than in Europe, so the opportunity exists for these countries to adopt legislative improvements already made in the European Union as well as embrace future reforms introduced in the EU. Indeed, a watching brief should be kept on progressive developments for farm animal welfare in *any* jurisdiction so that such improvements can be achieved on as wide a geographical basis as possible. There is scope also for greater international cooperation in developing standards of welfare for intensively farmed animals through the work of the World Organisation for Animal Health (OIE), and any such standards adopted by the OIE have to be consistent with the "Five Freedoms". In the achievement of reforms to improve the lives of intensively farmed animals throughout the world, the spirit and ideas of Ruth Harrison expressed in *Animal Machines* still have relevance.

Although considerable legislative progress had been made in the European Union towards improving the conditions of intensively farmed animals, Eurogroup for Animal Welfare (Eurogroup) produced in 2004 a comprehensive review of remaining concerns and proposed actions for further improving animal welfare in the EU, including that of farmed animals (Eurogroup for Animal Welfare 2004). This agenda provided a road map for the future, and it can be expected that with perseverance and political lobbying, further improvements based on incremental change will continue to be made to improve the lives of farm animals in Europe and eventually throughout the world. Since 2004 EU legislation has included the 2007 Council Directive prescribing minimum rules for the protection of chickens kept for meat production and the 2005 Council Regulations on the protection of animals during transport and related operations (Eurogroup for Animals 2008, pp. 11, 15). According to Peter Singer (2002, p. x), there was wide support throughout the European Union for the significant changes that had been achieved to improve farm animal welfare, and they have the support of leading experts in the EU on farm animal housing. Indeed, Singer regards the changes as a vindication of much of what animal advocates had been saying for a long time, going back to the pioneering work of Ruth Harrison in publishing *Animal Machines* in 1964.

By contrast to the position in Europe, improvements achieved for farm animal welfare in the EU were barely on the agenda in the USA (Singer 2002, pp. x–xi), even though a typical cage for laying hens in the USA provided only 48 cm^2 of space per hen and both calves and sows continued to be kept in narrow individual crates or stalls in which they could not turn around. Ironically, the greatest improvement for laying hens in the USA came from the restaurant company McDonalds that in the year 2000 announced it would require its egg suppliers to provide a 50 % increase in space to 72 cm^2 per hen that corresponded to the European standard at that time. In Australia, farm animal welfare standards are also behind those in Europe as evidenced in informative reports produced by the animal protection organisation Voiceless on pigs (Sherman et al. 2005) and on meat chickens (Sharman and Kossew 2008). Voiceless also produced a valuable report on food product labelling (Sharman 2007). During 2006 veterinarian Dr Hugh Wirth, President of RSPCA Australia and Past-President of the World Society for the Protection of Animals (WSPA), gave a televised address to the Australian National Press Club in Canberra entitled *Production Animal Welfare in the 21st Century* in which he observed (Wirth 2006, pp. 12–13) the current relevance of the 1965 Brambell Report that was being dusted off and looked at again after a long period in obscurity. Thus, the influence of Ruth Harrison and *Animal Machines* lives on.

In developing countries also there is enormous scope for improving the welfare of intensively farmed animals. Moreover, intensive animal agriculture presents a serious risk for the spread of disease, particularly in developing countries (World Society for the Protection of Animals 2005b, p. 12), and according to Cox (2007), the rapidly growing intensive animal agriculture in the developing countries of Latin America, Asia, and Africa is not only bad for animal welfare but contributes to the poverty problem in those countries. It is important to note that one of the guiding principles adopted by the World Organisation for Animal Health (OIE) is (Department of Agriculture, Fisheries and Forestry 2005, pp. 33–34):

> There is a critical relationship between animal health and animal welfare.

An important development for improving farm animal welfare on a global basis was the production by the World Society for the Protection of Animals (WSPA) in conjunction with Bristol University of a *Concepts in Animal Welfare* syllabus for use in veterinary education around the world (de Boo and Knight 2005, pp. 451–453). This syllabus comprises seven core and 23 elective modules. The modules specifically relevant to farm animals include a core module titled "Welfare assessment and the Five Freedoms" and two elective modules each titled "Farm animal welfare assessment and issues". It is highly desirable that future veterinarians all over the world be taught about animal welfare issues, and undoubtedly Ruth Harrison, with her emphasis on education, would have applauded this initiative of WSPA with which she was associated for many years.

Inspiration for the future is provided by the publication by Compassion in World Farming (CIWF) of reports showing high animal welfare systems in Europe suitable for farming laying hens (Arey 2004) and pigs (Arey and Brooke 2006). Indeed, there needs to be a change in human attitudes assisted by education as advocated by Ruth Harrison in *Animal Machines*, and this should embrace both producers and consumers so as to result in increased supply of and demand for animal products produced under high-welfare conditions. There is some evidence this is starting to happen, but much remains to be done.

Finally, it should be mentioned that a consideration of the influence and potential of *Animal Machines* that exposed the conditions prevailing in intensive animal agriculture involves not only the specific contents of the original work itself but also of other developments that have flowed from it. For instance, the publication of

Animal Machines led to the setting up of the Brambell Committee that stipulated five minimal freedoms for animals, and later these were expanded into what are now commonly known as the "Five Freedoms". These in turn have been influential in assessing animal welfare requirements and evaluating animal protection legislation, and they are included as a core component in the *Concepts in Animal Welfare* syllabus developed by WSPA in association with Bristol University for teaching animal welfare to future veterinarians around the world. Indeed, the journey of improvement in animal welfare involves the contributions of many people building upon one another, and it is the calculus of these contributions that leads through incremental improvements to enhanced welfare, changed attitudes, and a better life for farm animals. It was the publication of *Animal Machines* in 1964 that began this journey against the abuse of animals in intensive farming situations, and it is a journey that is still in progress.

3.5 Some Other Books

In 1975, Peter Singer published his celebrated book *Animal Liberation* (Singer 1975). The third chapter of that book entitled "Down on the Factory Farm" provides a revealing description of the suffering experienced by intensively farmed animals during their rearing, transport, and slaughter. The chapter deals with intensively farmed chickens (broilers and laying hens), pigs, calves (veal), and in the USA cattle feedlots. Moreover, the book contains a photographic gallery that includes disturbing images of intensively farmed animals. The book reinforces what had been exposed by Ruth Harrison about factory farming when she published her ground breaking work *Animal Machines* in 1964. In 1990, a revised edition of *Animal Liberation* was published (Singer 1990) in which Singer showed that, even though since 1975 there was greater awareness of animal welfare issues and attitudes were changing, the conditions and treatment of intensively farmed animals generally remained very bad.

Eating involves ethical issues, and in their book *The Ethics of What We Eat*, Peter Singer and Jim Mason (Singer and Mason 2006) compare the poor welfare treatment of animals that are intensively farmed to satisfy the standard American diet with the better welfare of animals that provide food for whom they call the conscientious omnivore. Also considered in the book is a vegan diet in which no animal products are consumed. Indeed, the book provides a relatively recent exposure of factory farming in the USA (including the secrecy that surrounds it) as well as more humane alternatives. Moreover, the book provides a guide to ethical eating that gives useful advice to a person wishing to contribute through his or her eating habits to a better life for animals, particularly those that are intensively farmed.

In 2008, a book *The Future of Animal Farming: Renewing the Ancient Contract* edited by Marian Dawkins and Roland Bonney (Dawkins and Bonney 2008) presents a deep level of debate in eloquently expressed views by a range of contributors that, among others, include Peter Singer, Bernard Rollin, Temple Grandin, and Colin Tudge. A review of the book by Sibley (2009, pp. 578–579) observes that the work will make readers think but implies that the reader will not come away with clear answers regarding the future of animal farming. Moreover, many contributions show the need for realistic and practical indicators of animal welfare for use in measuring progress or regression. In the foreword of the book, Peter Singer critically examines the current situation of farm animal welfare and draws an analogy with slavery, while Bernard Rollin argues pessimistically that the conflict in the USA between industrial agricultural production and animal husbandry will not be easy to settle. Temple Grandin shows from her own experiences that an individual can make a difference by concentrating on the achievable to produce rapid and effective results, while Colin Tudge sees a strategic approach rather than a pipedream as providing the solution to better welfare for farm animals.

After considerable research, award-winning American author Jonathan Safran Foer in 2009 published *Eating Animals* (Foer 2009) in which he exposes in a vivid way the conditions under which intensively farmed animals in the USA are reared, transported, and slaughtered. The book is important as it makes a valuable contribution to increasing transparency, and Foer will attract many readers not already familiar with animal welfare issues through his own personal following as an author. Even though Foer is in favour of vegetarianism for animal welfare and various other reasons, he advocated in an interview (Foer 2010) the moving away from emphasis on a person being "vegetarian or not" to one who "cares or does not", and he recognises that for the people who care, there are various options such as that of eating less meat that can make a valuable contribution to improved animal welfare and environmental protection.

Ruth Harrison in the conclusion to *Animal Machines* saw the need for a reassessment in basic human attitudes towards animals and advocated a role for education throughout the fabric of society. In the book *Education for Animal Welfare* (Eadie 2011) that forms part of the Springer Animal Welfare Series, the present author regards education in the broad sense of creating awareness and facilitating change. The book considers a number of educational themes relevant to improving animal welfare and examines various interrelated educational contexts through which the themes can be addressed. Both the themes and contexts have relevance to farm animal welfare. Among the educational themes considered are the issue of speciesism and the need for attitudinal change; the training of professionals, carers, and users involved with animals; and the scope for science to contribute, while the educational contexts examined include higher education and research; vocational and industry training; and animal protection organisations. It is hoped the book provides a useful framework with examples for developing the role of education in improving animal welfare that was regarded by Ruth Harrison in *Animal Machines* as so important for changing human attitudes.

3.6 Overview

Animal rights philosopher Tom Regan (1983) was entirely against the use of animals in agriculture, and animal rights lawyer Gary Francione (1995) regarded the use of legislation to protect animals as ineffective because of their status as property, although he acknowledged that absolute legal prohibitions can be used to provide some protection to animals. In any event, the continued farming of animals shows it is not likely to stop for some time at least, if ever, and their property status is unlikely to change in the foreseeable future.

The publication of *Animal Machines* by Ruth Harrison in 1964 was the catalyst for the movement against the advent of modern factory farming, and it led to the establishment in 1965 of the Brambell Committee in Britain and its recommendation of the five basic freedoms for farmed animals related to space requirements. These eventually developed into what are now known as the *Five Freedoms* for animals and cover all their behavioural needs. Moreover, the reading of *Animal Machines* had a huge influence on Peter Singer (2008, p. ix), who later played an important role also in exposing morally unacceptable conditions and practices in intensive farming situations.

The recommendations made by Ruth Harrison in *Animal Machines* in relation to veal calves, laying hens, and animal tethering have either been adopted or substantial adaptations made in Europe, but these have taken a long time to achieve. Indeed, it was only in 1997 that a Directive of the European Union imposed a ban on individual crates for calves to be implemented by 2007, in 1999 that a Directive on laying hens provided for a ban from 2012 on traditional battery cages that will be replaced by enriched cages, and in 2001 that a Directive on pigs banned the use of individual sow stalls from 2013 (except during early pregnancy), although in Britain such bans had been introduced prior to 1990 for veal calves and in 1998 for pregnant sows (Eurogroup for Animal Welfare 2003, p. 2; Singer 2002, p. x). In the USA progress in achieving improved conditions for intensively farmed animals has been even slower and less effective than in Europe (Singer 2002). Unless the World Organisation for Animal Health (OIE), that is officially guided by the *Five Freedoms* (Department of Agriculture, Fisheries and Forestry 2005, pp. 33–34), can develop internationally recognised and enforced standards for the satisfactory welfare of intensively farmed animals, there is a real danger that the conditions of a great many intensively farmed animals, including those in developing countries, will remain or become seriously impaired. Moreover, the current rules of the World Trade Organisation (WTO) prohibit import discrimination against farm products produced under less favourable animal welfare conditions than those existing domestically (Stevenson 2002), a matter Ruth Harrison was not able to anticipate in her recommendations in *Animal Machines*.

A major problem in tackling the poor welfare conditions experienced by millions of animals in factory farming situations is the powerful and secretive nature of the agricultural lobby and industry (Singer and Mason 2006; Foer 2009). Indeed, were it not for the writings and actions of Ruth Harrison (1964) and other animal activist organisations and individuals that followed her, the public generally and consumers of intensively farmed animal products would be totally oblivious of the conditions under which many of these are produced. Agricultural interests are not generally keen to argue the case for intensive farming publicly with animal welfare activists but rather lobby government behind closed doors. And usually governments are only too glad to support agricultural interests and get indirect credit for the availability of cheap agricultural products. It is only through the activities of animal activists that governments at times are exposed to criticism for allowing unacceptable animal welfare conditions. Indeed, many people are unaware that farm animals are treated differently under the law than companion animals (White 2007, p. 9; Eadie 2009, pp. 47–56), whereas the reality is that if a pet bird or a dog were kept under conditions allowed by the law for intensively farmed poultry and pigs, the owner would be prosecuted for animal cruelty.

In order to bring about more acceptable animal welfare conditions for intensively farmed animals as advocated by Ruth Harrison (1964), there needs to be proper transparency and widespread education so that the community generally and consumers in particular are made fully aware about the conditions under which intensively farmed animals are kept and treated. However, there needs to be action as well as awareness. With full transparency, coupled with enforced meaningful food product labelling, individuals would be able to make informed decisions about their consumption habits such as a choice to purchase free-range eggs or pork, and they could put pressure on their politicians as makers of the law to mandate improved conditions of welfare for intensively farmed animals. Morally, there is no proper place for a low price or political popularity that is based on animal suffering, and it is high time this is recognised more widely so that farmed animals can enjoy a pain-free and welfare satisfactory life in the process of bringing to humans farm products they demand if they are to continue doing so. Unfortunately, the quest Ruth Harrison began with the publication of *Animal Machines* in 1964 still has a long way to go.

Chapter 4
Immorality of the Treatment of Animals by Humans

Abstract The intensification of animal farming and use in laboratories in the latter part of the twentieth century led to much debate about the use of animals by humans. Peter Singer was one of the leaders in calling for a more humane approach to animal use, and his book Animal Liberation is widely hailed as the precursor to a rapidly growing activism movement. The varied approaches of Singer, Rollins, and others towards animal use are discussed, but it is recognised that Singer's book was most influential because it started an animal advocacy movement that has steadily grown since its publication in 1979.

4.1 Introduction

While a graduate student in philosophy at Oxford in the early 1970s, Peter Singer was first made aware of the suffering experienced by intensively farmed animals we eat through conversations with some of his philosophy colleagues (Singer 1975, p. 17). Moreover, after he began thinking about the ethics of the way we treat animals, the reading of Ruth Harrison's *Animal Machines* published in 1964 had a huge influence on him (Singer 2008, p. ix). In 1973 Peter Singer wrote a review of the book *Animals, Men and Morals* edited by Stanley and Roslind Godlovitch and John Harris that was published in *The New York Review of Books* (Singer 1973), and the enthusiastic response to that review provided the catalyst (Singer 1975, pp. 17–18) for writing the book:

Animal Liberation: Towards an End to Man's Inhumanity to Animals by Peter Singer published in 1975

E.N. Eadie, *Understanding Animal Welfare*, Animal Welfare 13,
DOI 10.1007/978-3-642-30577-1_4, © Springer-Verlag Berlin Heidelberg 2012

4.2 Contents of the Book

The Preface to the first edition of *Animal Liberation* (Singer 1975, p. i) begins with the words:

> This book is about the tyranny of human over nonhuman animals. This tyranny has caused and today is still causing an amount of pain and suffering that can only be compared with that which resulted from the centuries of tyranny by white humans over black humans. The struggle against this tyranny is a struggle as important as any of the moral and social issues that have been fought over in recent years.

Singer then invites his readers to reserve judgement until after reading the book to decide whether the opening paragraph is "a wild exaggeration or a sober estimate of a situation largely unknown to the general public".

In Chap. 1 of *Animal Liberation* (Singer 1975, pp. 21–43), Singer provides an introduction to philosophical and scientific aspects of equal consideration of interests, animal suffering, and discrimination by humans against animals known as speciesism. He observes that the capacity for suffering and enjoyment is a condition that has to be satisfied before interests can be attributed in a meaningful way. Moreover, the conclusions Singer argues for in the book follow on from the principle of minimising suffering. Chapters 2 and 3 of *Animal Liberation* (Singer 1975, pp. 44–100, 101–163) provide a practical exploration of two types of speciesism, namely the use by humans of animals in experimentation and animals reared for food. Singer recognises that suffering is inflicted by humans on other types of animal also, but he concentrates in the book on these two categories because of the enormous number of animals involved. Indeed, Singer observes that these activities "cause more suffering to a greater number of animals than anything else that humans do". Furthermore, animal experimentation is promoted by government and largely funded by taxes, while animal farming takes place because most people purchase and eat meat.

In Chap. 2, Singer (1975, pp. 44–100) describes the use of animals as tools for research and illustrates the enormous suffering inflicted on animals in laboratories operated by defence, industry, and university organisations. He advocates the immediate cessation of experiments that serve no direct or urgent purpose, and that in other cases, animal use should be replaced as quickly as possible by methods not involving animals. This assertion is supported by well-documented cases of seemingly unnecessary experiments then being conducted or performed over previous years. Singer (1975, pp. 79–80) gives a partial list by page number of the nature of the experiments described and the name of the organisation that paid for them. He acknowledges that measures such as the public exposure of what is happening behind closed doors, protesting against such experiments, and lobbying politicians may contribute to bringing about reform. However, he asserts that the problem is part of a larger problem, namely the need to eliminate speciesism, and he anticipates that one day our descendents will look back with horror and incredulity at what otherwise civilised people did to experimental animals, just as we now abhor past Roman gladiatorial atrocities and the eighteenth century slave trade.

In Chap. 3, Singer (1975, pp. 101–163) examines what happens to animals in factory farming situations and describes the suffering of various kinds inflicted on broiler chickens and layer hens, pigs, veal calves, dairy cows, and beef cattle during their rearing as well as when they are transported and slaughtered. His descriptions are not based on his own personal observations or on reports of people avowedly sympathetic to animal welfare but rather on sources likely to be positive to the farming industry, namely the magazines and trade journals of the industry itself. In *Animal Machines*, Ruth Harrison (1964) concluded that "cruelty is acknowledged only when profitability ceases", and Peter Singer (1975, p. 104) affirms that this attitude is exhibited in farming magazines both in the USA and Britain. Nevertheless, he observes that much can be learnt about the conditions of farm animals from such industry publications. Moreover, Singer (1975, pp. 137–147) discusses the findings of the 1965 Brambell Committee on intensive farming about which the public was largely ignorant, and he describes also other types of suffering such as dehorning, branding, and castration inflicted on traditionally farmed animals. The suffering of farm animals during transport and slaughter (including religious ritual slaughter) are discussed also by Singer (1975, pp. 149–159). Thus, in Chaps. 2 and 3, Singer provides an objective account of the speciesism found in the laboratory and on the farm, and this is reinforced by a powerful pictorial gallery of photographs with accompanying descriptions (photos 1–13) of animals used in laboratory and farming situations.

Chapters 4–6 of *Animal Liberation* (Singer 1975) deal with vegetarianism as a means of reducing animal suffering and human starvation concurrently (Chap. 4, pp. 164–189), provide a brief history of speciesism as exhibited in humans' dominion over animals (Chap. 5, pp. 190–216), and examine contemporary speciesism as reflected in defences, rationalisations, and objections to Animal Liberation (Chap. 6, pp. 217–250). In addition, *Animal Liberation* contains three useful appendices, namely cooking for vegetarianism (Appendix 1, pp. 251–267), additional reading (Appendix 2, pp. 268–269), and organisations in the USA and Britain working for change in attitudes to and the treatment of animals (Appendix 3, pp. 270–273). Throughout *Animal Liberation*, Peter Singer presents his evidence and arguments logically and without emotion, and the reader is left in a strong position to make a judgement as to the veracity of the opening paragraph of the Foreword quoted earlier regarding the manner in which animals are treated by humans as illustrated in Chaps. 2 and 3 in relation to experimental and farmed animals.

In 1990, a revised edition of *Animal Liberation* was published (Singer 1990). The most significant difference between the 1975 and 1990 editions of the book is the descriptions in Chaps. 2 and 3 (Singer 1990, pp. 25–94, 95–157) covering the treatment of experimental and farmed animals. These chapters have been brought up to date to document what was happening to animals currently in experimentation and farming and so avoid any claim that the treatment by humans of these types of animals described in the 1975 edition belonged only to the past. Nevertheless, since the publication of *Animal Liberation* in 1975, a modern animal movement had developed, and in the 1990 edition of *Animal Liberation*, Singer describes the achievements of the movement over the period since 1975. Indeed, the movement had become worldwide, and some important gains for animals had been achieved.

However, widespread animal exploitation continued, and a great deal remained to be done. Singer asserts that the movement should seek to achieve change through non-violent means in the tradition of Gandhi and Martin Luther King, not through violence that only breeds violence, and that the prospects of victory for the movement lie in the morality of its cause. In the 1990 edition of *Animal Liberation*, the ethical foundation of the arguments in the book have not been changed because these have withstood the test of time without any insurmountable objections through lectures, conference talks, philosophy seminars, and considerable discussion (both verbal and in print).

A more recent edition of *Animal Liberation* was published in 2002 (Singer 2002). This comprises the text of the 1990 edition but contains a new Preface that describes developments that have taken place since 1990, particularly in the treatment of intensively farmed animals in Europe (Singer 2002, p. x). These developments include the banning in Switzerland in 1991 of the battery cage system for producing eggs so that layer hens in Switzerland were able to scratch on a floor covered in straw or other organic material and lay their eggs in a sheltered and soft-floored nesting box. Moreover, the traditional bare wire cages for layer hens were to be phased out throughout the European Union (EU) by 2012. Furthermore, a ban on the keeping of veal calves in narrow individual crates, already in force in Britain prior to 1990, was to extend to all member countries in the EU by 2007, and the confining of pregnant sows in individual crates, banned in Britain in 1998, was to apply throughout the EU from 2013, except for the initial 4 weeks of pregnancy. However, similar changes had not occurred in the USA (Singer 2002, pp. x–xi), where the typical cage for a layer hen allowed only 48 in.^{2} of space per hen and both calves and sows continued to be kept in individual narrow crates or stalls. The only bright spot in the USA was the announcement by McDonalds in 2000 that it would require its egg suppliers to provide a 50 % increase above the traditional space allowance per hen, although this would result in only a small minority of American hens benefiting from the improvement. Singer refers also in the Preface of the 2002 edition of *Animal Liberation* (Singer 2002, pp. xii–xiii) to the recognition in the European Union of animals "as sentient beings" in the 1997 *Treaty of Amsterdam* Protocol on animal welfare in contrast to their classification as "agricultural products" in the original 1957 *Treaty of Rome*.

4.3 Influence of the Book

The publication by Peter Singer in 1975 of *Animal Liberation* provided the powerful foundation on which he has built a life of influence in the animal movement through his publication as author or editor of several other books, his academic contribution in teaching the moral philosophy of animal issues at various universities in different countries, his direct involvement in animal organisations and activism, particularly in Australia and the USA, and his widespread efforts internationally as a public intellectual giving voice to the cause of reducing animal suffering. Indeed, there are probably few, if any, other individuals who have been as influential as Peter Singer in drawing attention, directly or indirectly, to the immoral suffering inflicted by humans on animals. In this his influence has been profound.

It is suggested by Jasper and Nelkin (1992, pp. 37–38) that one measure of the growing intense interest in the new animal movement was the number of paperback copies of the book *Animal Liberation* in print. Indeed, they observe that Peter Singer's book was a "kind of Bible to the movement" and that many animal protection groups gave a copy of *Animal Liberation* to new members. Distribution of the book increased significantly in 1985 when 70,000 copies were in print, and by 1988 the figure had risen to more than 250,000. Then distribution fell off, possibly in anticipation of the revised edition of the book that appeared in hardcover in 1990.

In an article by Christine Townend (1999, pp. 8–9) paying tribute to the work of Peter Singer, she describes the dramatic importance of the book *Animal Liberation* in changing human perceptions about animals. For instance, at the time of its publication, media reports expressed amazement that people should become vegetarian because farm animals suffered to produce meat. Indeed, the public had no concept of the true conditions of farm animals. Following the formation of Animal Liberation (New South Wales) in 1976 under the influence of the book *Animal Liberation*, similar groups were established in every capital city of Australia, and they contributed to spreading the word about the way intensively farmed animals were treated. In an interview with Peter Singer by Animals Today in 1995 (Animals Today 1995, pp. 8–12), 20 years after the publication of *Animal Liberation*, Singer observed that animal rights had become an issue and there had been some major reforms for animals. These included the banning of battery cages for hens in Switzerland and the phasing out of factory farming in Sweden. In animal experimentation progress in developing alternatives to the LD50 and Draize tests had to be regarded as a major step forward. In Australia an important reform was the setting up of animal ethics committees with animal welfare representatives on them. Such ethics committees did not exist in 1975, and many animal experiments that were performed in 1975 with extreme animal suffering would not be approved in 1995. Moreover, the publication of *Animal Liberation* had a general influence on community awareness of animal suffering and on human attitudes towards animals. Singer observed, however, that while a great many people had been convinced about the evils of animal exploitation, many had not converted this recognition into a change in their behaviour. Thus, some progress was being achieved; there was a long way to go.

In addition to *Animal Liberation*, Peter Singer has authored or edited several other books dealing with animal issues. In 1981, James Mason and Peter Singer jointly published a book *Animal Factories: What Agribusiness Is Doing to the Family Farm, the Environment and Your Health* that described animal suffering in intensive farming situations, and in 1990, a revised edition of the book was published. Moreover, in 2006, Peter Singer and James Mason published *The Ethics of What We Eat* (Singer and Mason 2006) of which one part describes the conditions of farm animals used to provide the food consumed by those eating the traditional American diet, while other parts of the book deal with the more humane production of food for those the authors classify as conscientious omnivores and the diet of vegans who eat no animal products.

After attending an evening adult education course at New York University conducted by Peter Singer in 1974 based on the contents of what was to become the book *Animal Liberation* published the following year, Henry Spira devoted the rest of his life to campaigning for reforms against the abuse of experimental animals and farm animals, campaigns in which Peter Singer played an important role, and in 1998, Peter Singer published the biography of Henry Spira (Singer 1998), whom Singer has described (Singer 2006b, p. x) as the "most effective American campaigner for animals in the 1970s and 1980s". Based on the activities and experiences of Henry Spira described in the book, Singer (1998, pp. 184–192) outlines ten key points to assist future campaigners working to improve the lives of animals or indeed the lives of humans who are oppressed or exploited. These ten points provide a useful guide for the conduct of an effective campaign. Moreover, the successes of Spira showed that reforms can be achieved by means other than legislative change.

In 1985, Peter Singer edited a book *In Defence of Animals* (Singer 1985) that contained a selection of essays covering (1) The Ideas (that included, among others, essays by Tom Regan, Marian Stamp Dawkins, and Mary Midgley), (2) The Problems (that included essays by Richard Ryder and Jim Mason), and (3) The Activists and Their Strategies (including essays by Clive Hollands and Henry Spira). Peter Singer wrote the Prologue: Ethics and the New Animal Liberation Movement as well as an epilogue. In his prologue Singer described the book as providing a platform for the new animal liberation movement and concluded it with a plea for expanding the circle of ethics (to embrace animals). He regarded the essays in the book as showing how this could be done, both in theory and practice.

Twenty-one years later, Peter Singer edited a similar book entitled *In Defence of Animals: The Second Wave* (Singer 2006a, b) based on the same format, namely (1) The Ideas, (2) The Problems, and (3) Activists and Their Strategies. Singer wrote an introduction and a final word. In his introduction, Singer observes that the original edition of the book reflected the first generation of the modern animal movement, whereas the new edition represents the current state of the animal movement that had grown and matured over the previous 20 years. The essays in the new book include a repeat of the discussion by Marian Dawkins of the basis for assessing animal suffering, revised versions of three essays on animals in farms, laboratories, and zoos, and 14 entirely new essays. In reviewing the book, John Webster (2006, pp. 192–193) observes that a constant theme throughout the section on activists and their strategies is that it is more expedient and effective to work through the power of public opinion and consumer behaviour than attempt to achieve change through legislation, a matter on which he agrees. The part of the book Webster liked most was the essay entitled Ten Points for Activists by Henry Spira and Peter Singer (Singer 2006a, b, pp. 214–224) that he described as "sensible, charming, and wickedly effective". Moreover, Webster regarded the main target audience for the book as being young animal welfare activists seeking inspiration for their outrage and some direction in which to steer it.

Throughout his working life, Peter Singer has been involved professionally as an academic philosopher with appointments as Radcliffe Lecturer at University

College, Oxford, England 1971–1973, as Assistant Professor at New York University, USA 1973–1974, as Senior Lecturer at La Trobe University, Australia 1975–1976, as Professor of Philosophy at Monash University, Australia 1977–1999, as Ira W De Camp Professor of Bioethics at Princeton University, NJ in the USA since 1999, and as Laureate Professor at the Centre for Applied Philosophy and Public Ethics at the University of Melbourne, Australia since 2005, a post he holds jointly with his Princeton Professorship (Singh 2009, p. 1959). Each of these appointments has provided Singer with the opportunity to expound and spread his ideas about the ethics of the treatment of and suffering inflicted on animals by humans. Moreover, the teaching of moral philosophy, and of animal issues in particular, has become common in philosophy courses at universities in many countries (Singer 1990, pp. 241–242), and this was due in no small measure to the influence of Singer's book *Animal Liberation* published in 1975. Such teaching has been important in providing the moral underpinning for actions to enhance animal protection.

Singer and his book *Animal Liberation* have played substantial roles also in the establishment and operation of animal activist organisations, particularly in Australia and the USA. In Australia Christine Townend formed Animal Liberation (NSW) in 1976 after being inspired by Peter Singer and his animal liberation philosophy (Townend 1999, p. 9), and in 1980 she was co-founder with Peter Singer of the Australian Federation of Animal Societies (Townend 1981, pp. 109–120) that is now Animals Australia. Moreover, Peter Singer was Vice-President from 1984 to 1994 and President from 1994 to 1999 (until his appointment at Princeton) of the Australian and New Zealand Federation of Animal Societies (Singh 2009, p. 1959). In the USA the now major international animal protection organisation People for the Ethical Treatment of Animals (PETA) was formed in 1980 by Ingrid Newkirk and Alex Pacheco under the influence of Peter Singer's book *Animal Liberation* (Jasper and Nelkin 1992, pp. 29–30). In 1976, Ingrid Newkirk, who was born in Britain and raised in India, was appointed Director of Cruelty Investigations for the Washington Humane Society/SPCA in the USA. At that time, Alex Pacheco was at college in the USA when he read *Animal Liberation* then interrupted his studies to join the ship *Sea Shepherd* in the harassment of whalers. In 1980, Pacheco returned to the USA, met Ingrid Newkirk, and gave her a copy of *Animal Liberation*. Later that year, they jointly founded PETA.

Without detracting in any way from the achievements of Henry Spira as an exemplary campaigner for experimental and farm animals in the USA for 20 years or so from the mid-1970s, in a review of *Ethics into Action* (Singer 1998), the biography by Singer of Henry Spira, Glenys Oogjes (1998, p. 28) suggests that Singer played down his considerable influence on Spira both in awakening in him the plight of animals and in his subsequent highly successful activist activities. Peter Singer was instrumental also in establishing The Great Ape Project (Singer 1999, p. 10) that was formed in 1990 by a group of philosophers, scientists, lawyers, social scientists, and writers (among whom were Jane Goodall and Richard Dawkins) to bring about change in the moral status for great apes, namely chimpanzees, gorillas, and orangutans. Since 1993, Singer has been president of the Great Ape Project (Singh 2009, p. 1959).

The influence of *Animal Liberation* and the ideas contained in it have been enhanced greatly by the willingness of Peter Singer to be deeply involved as a public intellectual through his community speaking activities and media engagements of various kinds. Indeed, in 2005 he was named in the Time 100 Most Influential People and in the Top 100 Public Intellectuals by Foreign Policy/Prospect magazines (Singh 2009, p. 1959). It can be concluded confidently that Peter Singer through his book *Animal Liberation* as well as his other writings, academic involvements, activist activities, and public exposure have been monumental in changing human attitudes towards animals and about the manner in which they should be treated. However, there is still a long way to go in the realisation of an ethical world in which animals are treated with proper respect and their suffering inflicted by humans eliminated.

4.4 Potential for the Book

Animal Liberation continues to be widely cited, and it is likely the book will be read in the future by many more people. Moreover, the modern animal movement, in the development of which *Animal Liberation* has played an enormously important role, is likely to grow in strength and influence as more people become enlightened about the manner in which animals are treated by humans in the production of food. Indeed, more people will probably become vegetarian or vegan, particularly in Western countries, with resulting benefits for animal welfare, the environment, human health, and global equity. However, the eating of meat can be expected to continue and indeed increase in some parts of the world as people become more affluent and westernised. This will provide good reason for the messages in *Animal Liberation* to be spread more widely around the world so that there is improved welfare for the food animals that are consumed, and for various reasons, greater participation in the practice of vegetarianism and veganism in human eating habits is likely as more people become aware of the issues associated with meat consumption. In any event, whenever the opportunity arises, Peter Singer continues to speak out about animal suffering of all kinds inflicted by humans, and his influence remains strong.

4.5 Some Other Books

In 1981, *Animal Rights and Human Morality* (Rollin 1981) was published by Professor Bernard Rollin of the Colorado State University at which he holds joint chairs in Philosophy and in Physiology and Bioethics. The book is easy to read and inspirational. It makes a compelling case that humans should provide proper care and respect for the animals they use. The book contains four parts that cover (1) moral theory and animals, (2) animal rights and legal rights, (3) the use and abuse of animals in research, and (4) morality and pet animals. Rollin does not deal with food animals as they are covered in other works such as the then recently published book by James Mason and Peter Singer entitled *Animal Factories*.

An extensive coverage of ethical issues related to animals is contained in *The Animal Ethics Reader* (Armstrong and Botzler 2003) edited by Professors Susan Armstrong and Richard Botzler. The book brings together in a structured way a carefully selected and wide range of extracts, papers, and articles by various authors, including leading philosophers and scientists covering different perspectives on diverse issues relating to animal welfare and ethics. Each of the ten parts of the book deals with a separate topic and begins with an introduction by the editors outlining the nature of the items in that part. The book contains a splendid Foreword by Professor Bernard Rollin (2003, p. xiii) in which he observes that "this anthology is by far and away the best, most comprehensive, and readable introduction to a full range of issues associated with the new thinking and concern about animal treatment". In reviewing the book, Peter Thornton (2009, pp. 110–111) of the UK Home Office concurs with Rollin and enthusiastically recommends the book to a wide range of readers. The book contains a general introduction by the editors Armstrong and Botzler in which they provide a brief history of the moral thinking and treatment of humans towards animals. The first five parts of Armstrong and Botzler (2003) are of particular relevance to *Animal Liberation* (Singer 1975, 1990), and these cover: (1) theories of animal ethics; (2) animal capacities: pain, emotion, and consciousness; (3) primates and cetaceans; (4) animals for food; and (5) animal experimentation.

4.6 Overview

In *Animal Liberation*, Peter Singer provided powerful arguments against the infliction of suffering by humans on animals and illustrated the existence of discrimination by humans against animals known as speciesism by documenting the suffering inflicted on animals in the laboratory and on the factory farm. The widespread influence of Peter Singer has been facilitated by his clarity of argument, his capacity to communicate with a wide range of people, and his diverse activities as a writer, academic, activist, and public intellectual. Since the publication of *Animal Liberation*, there has been greater awareness in the community generally, particularly in Western countries, regarding the issue of animal suffering and increased disapproval of it. Nevertheless, such increased recognition and disgust need to be translated into action by means of personal consumer choice and the lobbying of government for reform, as well as influencing other people whenever the opportunity arises so that animal suffering is reduced or eliminated in practice. Indeed, awareness without action does nothing to improve the lives of animals.

Even though in *Animal Liberation* Singer concentrated on the suffering inflicted on animals in experimentation and testing and in factory farming because of the huge number of animals and the enormous total suffering involved in these activities, he recognised there are many other situations in which animals suffer at the hands of humans. Several such situations are listed by Eadie (2011, pp. 11–12), and these include, for instance, the slaughter of whales for alleged scientific research, the keeping of Asiatic black bears in small cages for extraction of their bile, the holding of pigeon shoots for recreational entertainment, and the bludgeoning to death of seal pups for their fur. Indeed, the infliction of suffering by humans on animals is so widespread and so diverse that there is need for a general change in culture and attitudes regarding the way humans treat animals to achieve extensive improvement in their welfare. This represents a huge challenge that can be assisted enormously by appropriate education, greater transparency, and widespread publicity. Moreover, it is important to expose specific types and individual cases of animal abuse and suffering wherever they occur, and in this the various animal protection organisations play a tremendously important role. Indeed, these organisations are essentially the free enterprise guardians of animal protection, and without them many types of animal suffering would go unnoticed. Fortunately, the activities of activist organisations and individuals do result in specific types of animal suffering being stopped as seen, for instance, in the banning in 2004 of hunting foxes with hounds in Britain and the tail docking of dogs for cosmetic purposes in Australia. Clearly, the problem of human-inflicted animal suffering has to be tackled both by bringing about change in culture and attitudes of humans generally and by focussing on specific situations involving suffering.

It needs to be recognised also that the problem is global, even though human attitudes to animals vary considerably from one part of the world to another as found by Phillips and McCulloch (2005, pp. 17–24). Further, there are many localised challenges in various parts of the world such as overcoming the long standing culture of bull fighting in Spain and the lack of sensitivity to the sentience of food animals often seen in markets in China as described by Dunlop (2008, p. 49). Such situations illustrate the difficult challenges that have to be faced if animals are to be protected as they should be. Moreover, there needs to be proper recognition throughout the world of the actual sentience of various types of animal over which humans have control or with which they interact. Indeed, there is probably no country where animals are fully and properly protected from human-inflicted suffering, although in some countries the situation is worse than in others.

Possibly, the World Organisation for Animal Health (OIE) could attempt to bring about a change in attitudes towards animal suffering in addition to its brief of establishing internationally recognised animal welfare standards (Department of Agriculture, Fisheries and Forestry 2005, pp. 33–34), and the adoption by the United Nations of a *Declaration on Animal Welfare* could help in this regard also. Indeed, one of the principles of the Manila *Proposal for a Declaration on Animal Welfare* agreed by the 19 participating countries (World Society for the Protection of Animals 2005a, pp. 3–4) is that "all appropriate steps shall be taken by nations to prevent cruelty to animals and to reduce their suffering". Further, the Preamble of the Declaration recognised that "animals are living, sentient beings and therefore deserve due consideration and respect". However, such admirable sentiments have to be converted into reality by effective communication and appropriate actions; otherwise, they mean nothing to the animals that are victims of human-inflicted suffering.

In *Animal Liberation*, Peter Singer argues in favour of vegetarianism and veganism as being a specially direct means of reducing the suffering inflicted by humans on the food animals we eat. Furthermore, the adoption of such practices has benefits also for human health, environmental protection, and international equity. Indeed, any adoption of vegetarian or vegan eating habits is to be applauded for animal welfare and other reasons. However, even though many more people are becoming vegetarian or vegan, and this has an immediate effect in reducing the animal suffering that would otherwise have occurred, there are many types of animal other than food animals that are subjected to suffering, and this is likely to continue unless human attitudes generally change or such abuses are prohibited. Moreover, the reality is that meat eating is increasing in most parts of the world even though many people are becoming vegetarian or vegan. Thus, there is a need for more people to become vegetarian and vegan or at least eat less meat, and it is important also that other forms of animal suffering are not overlooked. Indeed, they need to be exposed and addressed. And even when a particular form of animal suffering is stopped, whether by direct action or legislative means, there has to be eternal vigilance to ensure it does not recur.

Even though Peter Singer through the publication of *Animal Liberation* and other means has made an enormous contribution to making people more aware of human-inflicted animal suffering, provided a moral[1] basis for it to be stopped, and encouraged the adoption of vegetarianism or veganism as a direct means of reducing the suffering experienced by farm animals, there is a long way to go before a world free of human-inflicted animal suffering is realised. However, some encouragement can be drawn from the successes of other great social movements of the past relating to slavery, colonialism, racism, and sexism. Indeed, every effort possible using all available means should be made so that the battle against speciesism and animal suffering can justifiably be included on the list of socially enlightened advances.

[1] The elements of utilitarianism are outlined, which have had a major influence on how we evaluate our use of animals, for example in laboratories.

Chapter 5
History of Challenges to Speciesism

Abstract People have been studying human–animal interactions for very many years, and this chapter focuses on how our approach to animals has changed over time. Richard Ryder charted these changes elaborately in his book *Animal Revolution: Changing Attitudes Towards Speciesism.* His promotion of the term speciesism has provided a platform for continuing the social movement with a focus on animals; after equality for women, children, racial minorities, disabled people, and those with abnormal sexualities, it is logical to turn our attention to our treatment of animals. Ryder and others have been influential in describing the history of this movement, which inevitably helps to determine its current and future directions.

5.1 Introduction

The history of the animal movement is largely that of a battle against *speciesism*, a word coined in 1970 by psychologist Dr Richard Ryder to describe discrimination by humans against animals. The word is defined in the Oxford English Dictionary (Brown 1993, p. 2972) as:

> Discrimination against or exploitation of certain animal species by humans, based on an assumption of human superiority.

A history of the efforts by morally progressive humans to overcome the practice of speciesism in order to reduce animal suffering is described by Richard Ryder in his book:

Animal Revolution: Changing Attitudes Towards Speciesism by Richard Ryder published in 1989

E.N. Eadie, *Understanding Animal Welfare*, Animal Welfare 13,
DOI 10.1007/978-3-642-30577-1_5,

5.2 Contents of the Book

The book *Animal Revolution: Changing Attitudes Towards Speciesism* is an account of the history of the animal movement written by Richard Ryder, who had played a highly active personal role in developments in the animal movement over the previous three decades. This involvement extended from real grass roots activism to being President of the Royal Society for the Prevention of Cruelty to Animals (RSPCA) in Britain.

In the introduction to *Animal Revolution*, Ryder explains that the book is not a catalogue of cruelties or atrocities committed by humans against other species as these had been widely documented elsewhere, but rather it is an attempt to establish explanatory links for such behaviour as well as examine the evolving relationships between humans and non-humans and to use history to provide the basis for new ideas (Ryder 1989, p. 1). Ryder gives priority to the positive side of the human–non-human relationship as manifested in the writings and campaigns of those people concerned with protecting or improving the lives of animals. The book is divided into three parts of which Part 1 deals with The Past (Ryder 1989, pp. 15–177) that covers the period from early times to 1960, Part 11 examines Modern Times (Ryder 1989, pp. 179–306) from 1960 to the time of publication of the book in 1989, and Part 111 entitled Issues (Ryder 1989, pp. 307–336) provides a discussion of Speciesism, the term created by Ryder and now widely used. Even though Ryder concentrates in the book on the history of the animal movement in Britain, consideration is given also to developments in the movement in other countries and internationally.

Among the historical events discussed by Ryder in Part I of the book are the enactment in Britain in 1822 of the first ever national animal cruelty prevention legislation led by Richard Martin and Lord Erskine and the establishment in Britain 2 years later of a Society for the Prevention of Cruelty to Animals (SPCA) by the Reverend Arthur Broome, Lord Wilberforce, and others. The modern period since 1960 covered in Part II included the publication in 1964 by Ruth Harrison of her celebrated book *Animal Machines* that attacked cruelties inherent in certain types of "factory farming", the publication in 1975 by Peter Singer of his influential book *Animal Liberation* as well as the publication in 1975 of his own book *Victims of Science* dealing with laboratory animals, and the organisation in 1977 by the RSPCA of an Animal Rights Conference held at Trinity College, Cambridge, that was attended by most of the main figures in the animal movement at that time and concluded with *A Declaration Against Speciesism.* In 1980 both Eurogroup for Animal Welfare (Eurogroup) and the World Society for the Protection of Animals (WSPA) were formed, the latter following the merger of two existing international animal protection organisations. In Part III Ryder examined a range of issues related to speciesism. Moreover, he commented that gradually politicians had embraced environmental concerns of the public and that increasingly animal protection was seen as a major part of the "green" movement. Ryder concluded his book *Animal Revolution* with the observation (Ryder 1989, p. 336):

> People who are cruel to nonhumans are not all wicked; most are just unthinking. Those of us who seek change must not resort to hatred or violence, but press on with our campaigns to educate and legislate. We want people to open their eyes and to see the other animals as

> they really are – our kindred and our potential friends with whom we share a brief period of consciousness upon this planet.

Indeed, education of the public and pressure on politicians have been powerful forces in bringing about change to improve the lives of animals.

In 2000 a revised edition of *Animal Revolution* was published (Ryder 2000), and in it Ryder omitted parts of the first edition that had become outdated. Moreover, he added a new chapter entitled A Wider Perspective (Ryder 2000, pp. 195–221) that covered recent developments in ethical theory, the science of animal welfare, and the increasingly sophisticated political campaigns of the 1990s. This new chapter discusses, among other things, the European Union 1997 Treaty of Amsterdam Protocol on Animal Welfare and Protection that Ryder regarded as one of the most gratifying achievements for animal welfare in the 1990s. However, Ryder deals with the threat to improvements in animal welfare in Europe arising from the power of the World Trade Organisation (WTO) to sanction the erection of barriers to free trade. On the positive side Ryder describes the lead taken by Britain in the establishment of scientific farm animal welfare centres at the Universities of Bristol (Professor John Webster), Cambridge (Professor Don Broom), Edinburgh (Dr Natalie Waran and Dr Mike Appleby), and Oxford (Professor Marian Stamp Dawkins).

In a review of the revised edition of *Animal Revolution* (Ryder 2000), it is observed by Gauthier and Griffin (2001, pp. 108–110) that "it would be difficult to find a text that provides a more comprehensive history of man's changing use and relationship to non-human animals". However, the reviewers draw attention to the author being a person strongly against the use of animals and suggest this has had a colouring effect on his interpretation of historical facts. Nevertheless, they acknowledge that the book is not overly dominated by anti-speciesist arguments.

5.3 Influence of Book

In *Animal Revolution*, Dr Richard Ryder covers the political history of the animal movement, particularly in Britain but also in Europe, the USA, and internationally. In doing so, he provides not only a fascinating account written by a well-qualified insider but produces a very valuable case study embracing many useful lessons and ideas arising from the historical struggle to achieve improvement in the lives of animals. The work demonstrates the need for both perseverance and patience. Moreover, the book represents an appropriate tribute to the many people, who over the course of history have worked with great dedication to achieve gradual reforms that reduce animal suffering and create a better life for animals.

5.4 Potential for the Book

The book *Animal Revolution* enables the present state of the animal movement to be seen in historical perspective and shows that incremental improvements can be achieved, even if slowly. It can provide an inspiration for current and future workers in the movement in their efforts to achieve better animal protection. Moreover, the book contributes to greater understanding of and compassion for the relationship between humans and animals and presents a useful and practical case history of issues and problems associated with achieving change to improve the lives of animals.

5.5 Some Other Books

The development of the animal rights movement in the USA is described in *The Animal Rights Movement in America* by Finsen and Finsen (1994) in which they observe that the movement has had some impact, but aroused intense opposition from powerful industries whose vested interest was in maintaining the status quo. A split arose in the animal movement between reformists and abolitionists epitomised by the distinction between "larger cages" and "empty cages" (Armstrong and Botzler 2003, p. 10). In Australia the history of the RSPCA in South Australia established as an SPCA in 1875, around 50 years after the formation of the SPCA in Britain by the Reverend Arthur Broome and others, is described by Wallace Budd in *Hear the Other Side: The RSPCA in South Australia 1875–1988* (Budd 1988). The prefix "Royal" was granted to the SPCA in Britain by Queen Victoria in 1841 and to the SPCA in South Australia by King George VI in 1937 (Budd 1988, pp. 13, 94). The early history of the development of an animal liberation movement in Australia from the mid-1970s, influenced by the book *Animal Liberation* (Singer 1975), is described by Christine Townend in *A Voice for the Animals: How Animal Liberation Grew in Australia* (Townend 1981). She documents the formation of the Australian Federation of Animal Societies in 1980 with the involvement of both Peter Singer and Richard Ryder (Townend 1981, pp. 109–120). This federation is now called Animals Australia, and along with RSPCA Australia, it is one of the two major national animal protection organisations in Australia.

5.6 Overview

A knowledge of history provides an appreciation of how we have reached the stage we are at today in relation to providing some protection to animals against the way they are treated by humans and brings a realistic perspective on the situation. Many people assume that the mere exposure of suffering inflicted on animals by humans will be sufficient to make it stop. However, history has shown this is not the case even though such exposure is an important first step. Indeed, history has demonstrated that the road to improvement in the protection of animals against suffering has been long and hard, but that with perseverance and diligence improvements can be achieved, even though much remains to be done. It shows also the very important role animal protection organisations such as the British RSPCA have played in helping to create a better life for animals (Ryder 1989, 2000).

However, even though improved protection for animals has taken place in some more progressive jurisdictions such as in Britain, Switzerland, and more recently in the European Union, similar protection has not yet been achieved in many parts of the world. Furthermore, even in the jurisdictions that have led the way in creating a better world for animals, a great deal needs to be done before animals of all types, even though sentient beings capable of suffering, can be regarded as being properly protected. Indeed, the extent of the protection provided depends very much on the type of animal involved, whether companion animal, farm animal, experimental animal, or so-called pest animal (Eadie 2009, pp. 47–56).

History shows the need for patience and persistence in bringing about reform to give better protection for animals and provides a realistic appreciation of the difficulties involved and obstacles encountered. It shows also how improved protection in some jurisdictions has been brought about, indicates what remains to be done, and provides valuable lessons for the achievement of improvement in other jurisdictions where animal protection is less advanced or even absent altogether. Indeed, in some of the newly admitted countries of the European Union (EU) such as Latvia and Bulgaria (Eurogroup for Animal Welfare & RSPCA International 2001, pp. 44–47), the introduction of animal protection legislation only began in recent years, and on their admission to the EU in 2004 and 2007, respectively, these countries were required to adopt the *acquis communautaire* of EU law. In other jurisdictions such as in China, there is as yet no national animal protection legislation even though the terrible treatment of many types of animal in these countries shows that there is a great need for the introduction and enforcement of such legislation.

In relation to improvement in or introduction of animal protection legislation in countries where it is in its infancy or does not exist, there is no need to reinvent the wheel, and much can be gained from looking at the most suitable and successful provisions in legislation already adopted in progressive jurisdictions such as Sweden, New Zealand, and Britain (Department for Environment, Food and Rural Affairs 2004, p. 46) as well as in the European Union. Moreover, history has shown that in relation to human rights (Steiner and Alston 2000), there are now many international conventions designed to protect humans from discrimination and abuse, and in the future, similar types of international convention covering animals could provide better protection for them on a global basis (Eadie 2009, pp. 208–220). Indeed, history shows how improved morality and enlightened ideas have developed over the years to create a better world as seen by the abolition of slavery and reduced human discrimination on the grounds of race and gender, and a similar enlightenment appears to be developing at present in relation to the way humans treat animals. However, there is a very long way to go in this regard, particularly in countries where animals are hardly given a thought. Nevertheless, a knowledge of history as illustrated by Richard Ryder (1989) gives some reason for encouragement that with perseverance and enlightenment the future for animals and the way they are treated by humans will eventually be much brighter. Indeed, every effort possible needs to be made at present to help bring this about, and in this greater transparency, widespread education, changed attitudes, and compassionate concern have much to contribute in making it possible.

Chapter 6
Science of Animal Behaviour and Welfare

Abstract Following the major public interest in animal welfare in intensive farming and laboratory contexts, scientists became involved in the topic. A new discipline of Animal Welfare Science was born, led mainly by scientists in British universities. One of these, a veterinarian named John Webster, pioneered welfare studies in cattle and wrote a book about the problems that he encountered, called *Animal Welfare: a Cool Eye Towards Eden*. Such books help to bridge the gap between scientists and the public. They are read avidly by students of veterinary science and agriculture, giving them significant influence in the animal industries of the future. The books help to give the treatment of animals a firm foundation in science, rather than being based just on public sentiment.

6.1 Introduction

The establishment of animal welfare as a scientific discipline was regarded by Professor Don Broom of Cambridge University (Broom 2001, p. 25) as the most important general achievement for animals during the latter years of the twentieth century. A practical approach to the science of animal welfare and one that asks the animals what they want is contained in the book

Animal Welfare: A Cool Eye Towards Eden by John Webster published in 1995

6.2 Contents of the Book

When John Webster wrote *Animal Welfare: A Cool Eye Towards Eden* in 1995, he was Professor of Animal Husbandry in the Department of Veterinary Science at the University of Bristol in England, and he already had a distinguished career as a veterinarian and animal welfare academic behind him. In the book, Webster aims to provide constructive solutions in a pragmatic approach to humans' dominion over

E.N. Eadie, *Understanding Animal Welfare*, Animal Welfare 13,
DOI 10.1007/978-3-642-30577-1_6,

animals in which he recognises that animals are affected by what we do and not how we feel (Webster 1995, pp. viii–x). He analyses the nature of the mind, welfare, and suffering of animals to ascertain how they feel and then examines the major welfare problems faced by farm, wild, laboratory, and pet animals in order to advocate what we can do for them. His analysis is based on the logic of the "Five Freedoms", namely (Webster 1995, p. 11):

1. Freedom from thirst, hunger, and malnutrition
2. Freedom from discomfort
3. Freedom from pain, injury, and disease
4. Freedom to express normal behaviour
5. Freedom from fear and distress

Throughout the book, the approach of Webster (1995, p. 6) is to address, using scientific means, the nature of the mind and suffering of animals and then attempt to use this understanding to build practical recommendations for animal care and welfare that take moral considerations into account and are fair to both animals and humans.

The book *Animal Welfare: A Cool Eye Towards Eden* is divided into three parts. Part I (Webster 1995, pp. 1–16) provides an introduction to humans' dominion over animals and includes the role of the "Five Freedoms" in assessing the welfare of animals (Webster 1995, pp. 10–14). Animal welfare is defined in brief by Webster (1995, p. 11) as:

> The welfare of an animal is determined by its capacity to avoid suffering and sustain fitness.

His aim is to understand how animals view the world and then attempt to educate public opinion to perceive animal welfare as closely as possible to satisfy their needs. Part II of the book (Webster 1995, pp. 17–124) is an enquiry into the nature of animal welfare and animal suffering and contains an analysis of "how it is for them". Much of the evidence of what animals want is drawn from scientific research involving animals. For instance, in relation to an animal's sense of value, the pioneering animal behaviourist Marian Dawkins of Oxford University in her book *Through Our Eyes Only* published in 1993 states (Webster 1995, p. 22):

> If an animal will go to considerable lengths to make something appear or disappear then it is telling us by its actions that it values that commodity or its absence.

Webster (1995, p. 29) observes that pain is clearly one of the most severe sources of animal suffering. However, while being unhealthy is a powerful indicator of poor animal welfare, it is not necessarily a source of suffering. Moreover, suffering should not be equated with stress (Webster 1995, p. 38), but it occurs when the intensity or complexity of stresses goes beyond the capacity of an animal to cope or an animal is unable to take constructive action to avoid them. Webster observes that because many animals over which we have dominion have the capacity to suffer, we are presented with moral problems that require pragmatic solutions. A detailed analysis is made by Webster (1995, pp. 39–124) of the "Five Freedoms" that can be applied to any category of animal, and in the analysis, he makes reference by way of example to various types of animals.

In Part III of the book, Webster (1995, pp. 125–265) provides a constructive approach to problems arising from humans' dominion over animals and advocates solutions to "What we can do for them". This raises the question of how best to reconcile what we want from other animals with a proper and effective regard for their quality of life (Webster 1995, p. 127). In doing so Webster concentrates on the species and problems that he regards as requiring most urgent attention. Much of Part III deals with farm animals raised for food. First, their welfare generally is considered, and then a more detailed analysis is made of the welfare of pigs, poultry, cattle, and sheep. Moreover, Part III deals with the welfare problems of horses and pets, wild animals, and laboratory animals. Webster (1995, p. 265) observes that education needs to instil in people a proper sense of value for the life of sentient animals with the word "proper" implying a sense of balance. He regards the farming industry and scientific profession as having been "economical with the truth" and maintains that it should be possible to justify a system or procedure to "the reasonable man". If not, he feels it is reasonable to conclude that something *may* be wrong with the system or procedure. Moreover, Webster views humankind as having a capacity for compassion and asserts that when our immediate needs are met, we can afford to be compassionate.

Among the matters emerging from the analysis in Part III (Webster 1995, pp. 125–265) are:

1. Schemes such as The Royal Society for Prevention of Cruelty to Animals (RSPCA) "Freedom Food" that provide production quality assurances relating to animal welfare are based on a realistic approach for improving farm animal welfare (Webster 1995, p. 129). They avoid what Webster describes as "impossibilism", namely seeking too much and ending up with nothing. However, at the time of publication, such schemes influenced only a small minority of consumers.
2. During the latter half of the twentieth century, particularly during a period of around 20 years from the 1950s to the 1970s, there were greater changes in livestock production than over any other time in the history of agriculture (Webster 1995, p. 135), with livestock farmers taking the opportunity to expand production to meet unprecedented demand that was accompanied by reduced production costs made possible by removing animals from the land and enclosing them in buildings. However, the sentient animals involved were without a voice in the process, and economic forces failed to provide a mechanism for protecting their welfare.
3. It is possible to use the logic of the "Five Freedoms" to identify (Webster 1995, pp. 142–145) possible contributions to poor animal welfare linked directly to breeding, feeding, and housing of animals to achieve maximum productivity. Moreover, the move to intensification in livestock production has placed increased pressures on farmers as well as on their animals. The aim of Webster (1995, p. 145) is to improve the quality of life for all sentient animals on the farm, both farming families and their animals. Consumers have been the main beneficiary in the intensification of animal production, and there is need for consumers to concede more to ensure farm animals get a fair deal. Indeed, if the lives of animals are to have more value, we will have to value them more highly.

4. The production of food products from pigs and poultry raises the most concern on welfare grounds, and it is these products the public has increasingly chosen to buy (Webster 1995, p. 146). Accordingly, Webster examines in some detail the welfare problems of commercial production systems involving pigs and poultry (Webster 1995, pp. 146–166), and he proposes priorities for change to improve animal welfare in ways consistent with good and economically sound husbandry.
5. In relation to broiler chickens, genetic selection for rapid growth and accentuation of breast muscle has created "leg weakness", a euphemism for a series of pathological conditions involving bones, joints, and skin (Webster 1995, p. 156), arising from birds growing too heavy for their limbs and/or becoming so distorted in shape that unnatural stresses are imposed on their joints. Indeed, about one in four birds in heavy strains of meat chicken and turkey experience chronic pain for about a third of their lives; so bearing in mind that poultry meat consumption in the UK alone was in excess of a million tonnes a year, Webster regarded this problem in terms of both magnitude and severity as constituting the single severest systematic exemplification of the inhumanity of humans towards another sentient animal.
6. In the opinion of Webster (1995, pp. 157–158), the battery cage for laying hens in its then existing form did not meet acceptable minimum standards for layer hen welfare, and he advocated what he regarded as essential features for an acceptable cage for such birds, namely the provision of suitable nesting boxes, a perch, and 900 cm^2 of space per bird. In reviewing *Animal Welfare: A Cool Eye Towards Eden*, Ruth Harrison (1996, p. 454) acknowledged that it is difficult to incorporate litter into modified cages for layer hens but observed that evidence shows from the perspective of a hen that a cage needs to contain litter to be acceptable in welfare terms.
7. In relation to pigs, Webster (1995, p. 165) is satisfied; there is sufficient evidence to justify a ban, as was to take effect in the UK from 1999, on the housing of dry sows in individual unbedded stalls, with or without tethers. Indeed, when all the evidence including that of chronic discomfort, pain, and injury in addition to abnormal stereotypic behaviour are considered, Webster regards the case for a ban on the semi-permanent housing of dry sows in unbedded stalls as overwhelming.
8. According to Webster (1995, pp. 168–169), severity of suffering has to be considered in terms of both its intensity and duration. For instance, the intensity of suffering of a cow during Halal or Shekita religious slaughter is extremely severe, but the duration is brief, whereas the intensity of suffering of a battery hen is less severe but lasts throughout the adult life of the hen. Webster (1995, p. 169) believes the suffering of a hen in a traditional battery cage greatly exceeds that of a beef cow ritually slaughtered after living a life at grass. Nevertheless, just as he regards traditional battery cages for hens as unacceptable (Webster 1995, pp. 157–158), he would probably agree with Egerton (1992, pp. 47–49) that the slaughtering of a conscious animal should be banned.

Indeed, the Moslem community in Australia has agreed that Halal killing using head-only pre-slaughter stunning is in accordance with the Quran (Wirth 2005).

9. An undoubted source of suffering in dairy cows is that caused by pain associated with lameness, particularly injuries to feet (Webster 1995, pp. 172–175). Indeed, surveys of foot lameness in dairy cows in the UK reveals an incidence of around 25 % a year, so it is a serious and painful problem for cows that have an average working life of 4 years. Webster regards education involving the application of existing knowledge as the most effective means for reducing such lameness in cows.
10. One of the most strange and cruel forms of livestock production is the rearing of calves in small wooden crates and feeding them an iron-deficient diet to produce white veal (Webster 1995, pp. 186–191). In contrast to pig and poultry production to satisfy widespread demand for cheap meat, the white veal industry has evolved to supply an expensive luxury food market catering for a minority of people. Generally the production of veal abused each of the "Five Freedoms" for animals and was associated with the export from the UK of hundreds of thousands of calves a year, particularly to Holland and France. As an alternative to this, research at Bristol University has demonstrated it is possible to rear healthy calves under good welfare conditions to produce meat acceptable to a veal butcher.
11. In relation to the live export trade in calves and sheep, Webster (1995, pp. 194–196) applauded Compassion in World Farming (CIWF), The Royal Society for the Prevention of Cruelty to Animals (RSPCA), and others for their outrage that he regarded as necessary to get things done. Many calves and sheep suffer during long-distance transport, although few actually die. The main welfare problems they experience during transport, particularly by road, are hunger, thirst, heat, cold, and exhaustion. Webster (1995, pp. 195–196) argues that legislation to improve the welfare of calves and sheep during transport should be based on a "stick and carrot" approach, with continuous journey lengths being restricted for conventional vehicles and incentives offered to transporters to provide vehicles and journey plans that allow animals to travel without suffering or the need for interrupted journeys.
12. It is observed by Webster (1995, p. 197) that the majority of stresses experienced by cattle and sheep during their last day of life are those associated with handling, transport, and lairage prior to their stunning, particularly in the case of animals unfamiliar with handling procedures. Accordingly, Webster recommended that the UK slaughter legislation provision that "animals must be rendered instantaneously insensible to pain" should be amended to read "at the place of slaughter animals must be handled, rendered unconscious and killed in such a way as to minimise pain and suffering". Such a change would broaden the legislation to protect animals during the entire slaughtering process, both prior to and during their actual killing.
13. The enormous scale of farm animal production is provided by statistics given by Webster (1995, pp. 127, 233). In the UK alone, the average human omnivore by the age of 70 consumes 550 poultry, 36 pigs, 36 sheep, and 8 oxen as well as

10,000 eggs and dairy products equivalent to 18 tonnes of milk. By contrast, about four laboratory animals, half of which are mice, are used in scientific procedures for each human over a similar lifetime. Moreover, more than 40 % of European farm incomes are subsidised by taxation revenue, yet no attempt had been made to apply subsidies as a direct incentive to improve animal welfare. Increasingly, public opinion was calling for better animal welfare, and Webster believed that with proper support farmers could provide it.

In 2005 a new book by John Webster entitled *Animal Welfare: Limping Towards Eden* (Webster 2005) subtitled "A practical approach to redressing the problem of our dominion over the animals" was published. Webster regarded it as a separate book to *Animal Welfare: A Cool Eye Towards Eden* (Webster 1995) subtitled "A constructive approach to the problem of man's dominion over animals", although *Limping* could be regarded as a sequel to or revised edition of *A Cool Eye*. In any event, Webster recommends that both books should be read (Webster 2005, p. x).

During the 10-year period following the publication of *A Cool Eye* (Webster 2005), there was an enormous increase in active concern regarding animal welfare, and even though less marked, there were some impressive constructive actions manifested in new legislation, new codes of practice for farm animals, new codes of ethics for laboratory animals, and new developments in quality assurance schemes for high-welfare food production (Webster 2005, p. x). These developments were supported by new research findings and a greater understanding of animal welfare science. *Limping* (Webster 2005) is a review of this progress on a journey involving improved animal welfare that Webster regards as a journey without end. In terms of structure, *Limping* comprises 11 chapters of which the first four deal with general issues of animal husbandry and welfare; the next six chapters cover specific types of animal use including animals for food (pigs, poultry, cattle, and other ruminants as well as animal transport and slaughter), animals for science and biotechnology, animals in sport, and animals as pets in which the *Five Freedoms* for animals are used to identify welfare problems; and the final chapter examines the way forward including a review of the broader issues, important current concerns, welfare assurance programmes, and education. Even though *Limping* contains some technical matters, it is generally easy to read and is practical. Indeed, in his review of the book, David Fraser (2005, pp. 391–392) concludes that "almost anyone involved in animal welfare policy and reform would profit from reading it".

6.3 Influence of the Book

When *Animal Welfare: A Cool Eye Towards Eden* (Webster 1995) was published in 1995, John Webster was Professor of Animal Husbandry in the School of Veterinary Science at the University of Bristol, and he was already highly regarded as a veterinarian, academic, and reformer. Indeed, through his membership of the Farm Animal Welfare Council (FAWC) in Britain, Webster played an important role in developing the *Five Freedoms* for animals, so they covered the entire range of animal needs, not merely space requirements as advocated in the Brambell Report (Webster 1995, p. 11). In both *A Cool Eye* and *Limping*, Webster uses the *Five Freedoms* as the basis for assessing animal welfare, and this provides a meaningful and practical approach to understanding and improving animal welfare. It is difficult to separate the influence of the book *A Cool Eye* published in 1995 from the influence of Webster as an academic and reformer. Nevertheless, some indication of the impact of *A Cool Eye* is given by Fraser (2005, p. 391) in his review of *Limping* in which he describes *A Cool Eye* as a "well received book". Moreover, Webster's more recent book *Limping* published in 2005 has continued and extended the influence of *A Cool Eye*.

The influence of both *A Cool Eye* and *Limping* will have been enhanced by them being reasonably easy to read by a wide audience that includes not only educators and students in the animal sciences but others with an interest in improving animal welfare such as veterinary and animal scientists, animal activists, law reformers, quality assurance developers, and indeed anyone involved directly with animals such as livestock farmers, animal transporters and slaughterers, experimental animal carers and scientists, and pet owners. This is facilitated even more by the general chapters being followed by chapters dealing with specific categories of animals. Moreover, the analysis of animal welfare issues is made systematic by continued reference to the *Five Freedoms* as a means for assessing and improving welfare. The usefulness of both *A Cool Eye* and *Limping* is accentuated also by the books drawing attention to specific issues in need of reform as well as offering possible solutions to the problems. Such issues included the use of crates for veal calves, battery cages for laying hens, lameness in dairy cows, and the handling of animals prior to slaughter. Indeed, reforms have taken place already or are in progress in some jurisdictions, particularly in Europe, in relation to the abolition of veal calf crates, sow stalls, and traditional cages for egg-producing hens. In these developments, John Webster through *A Cool Eye* and his other activities as an academic and veterinarian will have had considerable influence. Moreover, both *A Cool Eye* and *Limping* discuss the role of quality assurance schemes and of animal welfare education as means of creating a better life for animals, and in each of these matters, the University of Bristol has been particularly active and influential.

6.4 Potential for the Book

Both *Animal Welfare: A Cool Eye Towards Eden* (Webster 1995) and its successor *Animal Welfare: Limping Towards Eden* (Webster 2005) are considered together. These books are informative and practical in their approach to improving animal welfare that is greatly needed because of the suffering imposed by humans' use of his dominion over animals. The potential for the books is accentuated by the wide audience that can benefit from reading them. This audience includes:

1. Teachers and students of animal and veterinary science for whom an appreciation of animal welfare issues is important
2. Animal activists whose credibility can be enhanced by a sound knowledge of animal welfare science
3. Law reformers including bureaucrats and politicians who need to be properly informed about the issue of animal suffering, presented with scientific evidence about situations involving suffering where the law requires change, and about possible solutions to problems that have been identified
4. Developers and implementers of quality assurance schemes that have to be informed by animal welfare science to make them credible and genuinely improve animal welfare
5. Anyone involved in the farming, transport, and slaughter of animals so they have an understanding of animal welfare issues and can gain valuable knowledge from reading sections of the book relevant to their specific activities
6. Animal carers, experimental scientists, and animal ethics committee members involved with laboratory animals so they have a proper appreciation of animal welfare
7. Pet owners who can enhance their knowledge of animal welfare issues as well as of the needs of their pet
8. Consumers generally so they can make more informed decisions about what they buy or do not buy when considering purchases involving animal products
9. Members of the media so they can be better informed about animals generally and be inspired to cover issues where improved animal protection is needed
10. School teacher trainers and teachers themselves so as to encourage the incorporation of animal welfare issues into the curriculum, particularly as young people are particularly receptive to change

Even though *A Cool Eye* and *Limping* are in part quite technical, they each contain much that is easy to read and they enable a good appreciation of animal welfare issues. Indeed, the books have much more to offer about animal welfare than something that is simplistic and superficial. Indeed, they provide a good introduction to any reader wishing to go more deeply later into any particular aspect of the subject.

Education is regarded as important in both *A Cool Eye* and *Limping*, and Webster (2005, pp. 267–269) advocates that education needs to be pursued at three levels, namely increasing awareness, education for the public good, and education of the professional. In relation to animal welfare education for students in veterinary science as well as other vocations involving animals, the World Society for the Protection of Animals (WSPA) in conjunction with Bristol University has developed a flexible *Concepts in Animal Welfare* syllabus (Webster 2005, pp. 269–271; de Boo and Knight 2005, pp. 451–453) that is computer based and is making a significant contribution to animal welfare teaching worldwide, as anticipated by Webster. Indeed, the course has the potential to make an enormous contribution in the future towards improving animal welfare throughout the world, and in this animals will be the beneficiaries.

Another matter addressed by Webster (1995, pp. 129–130, 2005, pp. 78–96, 264–267) is the development, implementation, and assessment of quality assurance schemes to improve animal husbandry and welfare on the farm. Of these, the first was the British RSPCA "Freedom Food" scheme set up in 1994, and such schemes can be used by retail food suppliers and fast food outlets to ensure the proper welfare of farm animals obtained from their suppliers. The schemes need to identify animal welfare problems, the cause of the problems, and actions needed to remedy them (Webster 2005, pp. 93, 265–266). Moreover, customers must be prepared to trust the assurances of quality given. Indeed, trust in a high-welfare quality assurance scheme requires credibility of the entire quality control package that includes compliance with good husbandry provisions, independent assessment of welfare outcomes, and enforcement of action to resolve any problems. Finally, for a scheme to be successful, all parties need to see benefit from it, whether in terms of financial gain or improved animal welfare. *A Cool Eye* and *Limping* have the potential to influence future expansion in quality assurance schemes to provide improved welfare for various types of food animal in a range of jurisdictions around the world. Moreover, Bristol University could well play a direct role in this.

In *Limping* (Webster 2005), a number of remaining serious animal welfare issues are addressed. These include:

1. Exemptions from the British slaughter regulations made on religious grounds that allow a conscious animal to be killed without prior stunning during Halal and Shekita slaughter (Webster 2005, p. 257). This results in suffering that is avoidable. Similar exemptions apply in other jurisdictions also.
2. Allowance of the continued import into the European Union (EU) of eggs produced by hens kept in battery cages after such cages are banned in the EU from 2012 (Webster 2005, p. 257). The legislative resolution of this problem is hampered by World Trade Organisation (WTO) rules, but it represents an anomaly in the fully effective achievement of improved animal welfare that needs to be rectified (Stevenson 2002; Eadie 2009, pp. 149–154).
3. The use of genetic manipulation of animals that directly results in animal suffering (Webster 2005, pp. 155–156, 257–259). Cases of such genetic engineering include the breeding of Belgian Blue cattle whose muscle development requires surgical intervention involving Caesarean section in order to calve and the breeding of poultry, including broiler chickens and turkeys, that are so heavy or distorted in shape they are unable to stand up without experiencing pain in their leg joints.

By drawing attention to such practices, it is more likely that action will be taken in the future to abolish them, as has been the case in relation to veal crates, sow stalls, and traditional cages for laying hens, even though the progress of change regarding these has been slow. Indeed, *Limping* has the potential to be influential in that it provides an authoritative voice for reform, and it can give a credible stimulus to animal activists to play their part in bringing about change.

The books *A Cool Eye* and *Limping* (Webster 1995, 2005) contain a wealth of authoritative information and ideas on animal welfare science that are accessible to a wide readership, and they have the potential to make an enormous contribution when applied to improving the lives of animals of many types, notwithstanding the fact that animal production systems change quite rapidly and any book on animal welfare becomes dated quickly. In the final chapter of *Limping*, Webster (2005, pp. 249–271) gives a valuable view of the road ahead on a journey to enhance the welfare of our fellow creatures over whom we have dominion, and even though the journey will be long, hard, and never ending, it will always be necessary to take the next step.

6.5 Some Other Books

In 1990, *Farm Animal Behaviour and Welfare* by A.F. Fraser, Professor of Surgery (Veterinary) at the Memorial University of Newfoundland, Canada, and D.M. Broom, Colleen Macleod Professor of Animal Welfare in the Department of Clinical Veterinary Medicine at the University of Cambridge, England, was published (Fraser and Broom 1990). It is described as the third edition of *Farm Animal Behaviour* by A.F. Fraser, of which the first and second editions were published in 1974 and 1980. The book by Fraser and Broom (1990) is a comprehensive guide to the behaviour of farm animals and to the objective evaluation of their welfare based on scientific evidence. It provides practical information for people involved with animal farming and veterinary practice, and it reviews available scientific information regarding animal welfare assessment including the effect on animals of various management methods and housing conditions (Fraser and Broom 1990, pp. vii–viii). In addition to much new material on animal behaviour not included in the second edition of the book (Fraser 1980), there is a substantial final section in the third edition comprising 12 chapters on farm animal welfare (Fraser and Broom 1990, pp. 256–384) that is entirely new. This section on animal welfare contains much practical information, and it includes a comparison of the welfare of cattle, pigs, and poultry under different systems. It reflects the pioneering work on animal welfare by co-author Don Broom, who on his appointment at Cambridge University in 1986 became the first ever Professor of Animal Welfare in the world. Since its publication in 1990, the book has been the major student text on animal welfare (Phillips 2011a). In 2007, a fourth edition of the book entitled *Domestic Animal Behaviour and Welfare* was published (Broom and Fraser 2007), and as the title suggests, it is expanded to include companion animals. Moreover, the book is updated to take into account developments over the previous 10 years. The book provides a valuable recent text for students and practitioners in the veterinary and agricultural sciences.

A very readable book on animal welfare entitled *What Should We Do About Animal Welfare?* by Dr Mike Appleby, Senior Lecturer in Farm Animal Behaviour at the University of Edinburgh, was published in 1999 (Appleby 1999). The writing of the book was supported by a Hume Fellowship awarded by the Universities Federation for Animal Welfare (UFAW) that aims to enhance the welfare of animals by improving our ability to understand them. The book provides a lucid and balanced account of the science and ethics of animal welfare and covers the more important problems involved in the use and abuse of animals. It explains why these have presented difficulties and how progress can be made in the future. The book provides a good introduction to anyone interested in animal welfare including students, researchers, veterinarians, farmers, and members of the community generally. The book comprises nine chapters as well as an appendix containing a list of useful addresses. Each chapter of the book is on a separate topic relevant to animal welfare, and each contains a valuable list of conclusions and references. Moreover, the book is brought to life even more by the inclusion of a number of illustrative sketches. In her review of the book, Ruth Harrison (2000, pp. 92–94) described it as being well written and absorbing, both thoughtful and thought provoking, and a valuable read for individuals, discussion groups, and decision-makers at every level. The book was reprinted in 2005.

An informative collection of recent books on animal welfare is the *Springer Animal Welfare Series* that comprises twelve separate books as listed below:

Volume 1: The Welfare of Horses (Waran 2002)
Volume 2: The Welfare of Laboratory Animals (Kaliste 2004)
Volume 3: The Welfare of Cats (Rochlitz 2005)
Volume 4: The Welfare of Dogs (Stafford 2006)
Volume 5: The Welfare of Cattle (Rushen, de Passillé, von Keyserlingk, and Weary 2008)
Volume 6: The Welfare of Sheep (Dwyer 2008)
Volume 7: The Welfare of Pigs (Marchant-Forde 2009)
Volume 8: The Welfare of Animals: The Silent Majority (Phillips 2009)
Volume 9: The Welfare of Domestic Fowl and Other Captive Birds (Duncan and Hawkins 2010)
Volume 10: Education for Animal Welfare (Eadie 2011)
Volume 11: The Welfare of Farmed Flightless Birds (Glatz, Lunam and Malecki 2011)
Volume 12: Funding and Diplomacy in Animal Welfare Organisations (Roeder 2011)

It is seen that nine of the books in the series deal with the welfare of a specific species or group of animals. In each case, welfare is considered in relation to an animal's needs with concentration being given to nutrition, behaviour, reproduction, and the physical and social environment (Phillips 2011a, p. vi). The other two books in the series deal with the welfare of animals generally and with education for animal welfare. This volume represents the latest in the series. The Series Editor is Clive Phillips, Foundation Professor of Animal Welfare and Director of the Centre for Animal Welfare and Ethics (CAWE) in the School of Veterinary Science at the University of Queensland, Australia. In relation to farm animals covered in the series, it is observed by Phillips (2008, p. 1):

> It is hoped that these books will provide inspiration to all those people who are working tirelessly to improve animal welfare with little thanks or praise .... Working *for* animals, as opposed to *with* animals, requires courage and the people involved often face criticisms

from those who seek to get the maximum from their animals for the smallest input. Some whose livelihoods depend on using animals are averse to any change that will affect the short-term profitability of their enterprise, although by improving animal welfare they are guaranteeing their future markets and ethical acceptability of their business in the long-term.

Moreover, each of the books in the series is designed to contribute towards a culture of respect for animals and their welfare (Phillips 2011a, pp. v–vii), and the series provides an authoritative set of texts suitable for researchers, lecturers, students, and practitioners.

In 2010, *Improving Animal Welfare: A Practical Approach* edited by Dr Temple Grandin, Department of Animal Sciences at the Colorado State University in the USA, was published (Grandin 2010). The book aims at bridging the gap between scientific research and practical application and provides a guide to the practical evaluation and auditing of welfare problems with emphasis on the importance of measuring conditions that compromise welfare. The book comprises 15 chapters covering a range of topics and includes contributions by authors from the USA, Canada, UK, and New Zealand. In a review of the book, Andrew Fisher (2010, p. 546) observes, not surprisingly for a text edited by Temple Grandin, that emphasis is placed on providing practical tools for enhancing welfare and the meeting of welfare standards in animal production, transport, and at slaughter. Moreover, coverage in the book focuses on the major production and draught species, namely cattle, pigs, poultry, sheep, horses, and donkeys. The book provides a useful reference for those wishing to assess and ensure animal welfare in their own operations or as the basis for developing a welfare assurance programme. It could also supplement the teaching of production animal welfare to students in the animal and veterinary sciences.

6.6 Overview

The science of animal welfare, as in *Animal Welfare: A Cool Eye Towards Eden* (Webster 1995), provides a rigorous basis for understanding the needs of animals and for improving their protection against suffering inflicted by humans. Such suffering often arises from ignorance or lack of thought regarding the appropriate care required to satisfy the proper welfare needs of an animal. Consequently, it is important for scientific discoveries and knowledge about the nature and needs of animals to be communicated to a wide audience, particularly those involved directly or even indirectly with animals, in order for them to be fully effective and their potential value translated into improved protection for animals. Indeed, the findings of animal welfare science have relevance to many categories of people including:

1. Farmers, transporters, and slaughterers of animals
2. Practitioners, lecturers, and students in the animal and veterinary sciences
3. Animal activists and members of animal protection organisations

4. Bureaucrats, politicians, and lawyers involved in animal law reform and legal practice
5. Developers and operators of quality assurance schemes
6. Members of the media as informers of the public
7. Consumers of animal products
8. Pet suppliers and owners
9. Laboratory animal carers, researchers, and animal ethics committee members
10. Teacher educators and school teachers with the potential to influence young people
11. Zoo operators and keepers
12. Officials of the World Organisation for Animal Health (OIE) involved in developing animal welfare standards

It is important also that anyone involved with animals has a good appreciation of the *Five Freedoms* for animals. These can be used to assess the welfare of a particular animal species or to evaluate the veracity of a legal provision designed to protect animals.

In various countries, particularly in Europe and the USA, huge subsidy payments from taxation revenue have been made over the years to support animal agricultural production, and these have contributed to an abundance of cheap food without full regard for animal welfare. It is time similar subsidies were directed towards improving animal welfare. Indeed, such payments could be tied to appropriate improvements in animal welfare. This may well now be politically acceptable as people are becoming increasingly concerned about animal welfare issues. Moreover, the government could give direct support to the development, implementation, and auditing of animal welfare quality assurance schemes for food products. It could also introduce effective mandatory and properly enforced labelling of animal products so that consumers are able to make informed decisions related to the conditions under which the animal products they purchase are produced.

Animal welfare is an international issue, and it is important that animal welfare science is understood in both developed and developing countries. In this regard the *Concepts in Animal Welfare* syllabus developed by the World Society for the Protection of Animals (WSPA) in conjunction with Bristol University is playing a valuable role. It is important also that the World Organisation for Animal Health (OIE) that is developing animal welfare standards for global use gives proper regard to the *Five Freedoms* as specified as one of its guiding principles. Moreover, if any standards developed by OIE are no more than the "lowest common denominator", it is important that this does not have the effect of lowering standards in some countries by making the OIE standards a widely accepted benchmark. Indeed, the aim should be for the OIE standards to be progressively improved so that animal welfare all over the world satisfies the proper needs of the animals we control to produce food or satisfy other human requirements. In any event the OIE in its animal welfare activities should never become just a front for agricultural production interests rather than for proper animal protection.

The rapidly developing science of animal welfare has a great deal to contribute in the creation of a better life for animals through increased knowledge and understanding of their needs, and its findings have to be effectively and widely communicated so their potential to improve the welfare of animals all over the world is properly realised. Indeed, animals know of no man-made geographical boundaries, and their capacity to suffer is universal. The great challenge is to spread the message of animal welfare throughout a world of diverse cultures and political systems, and in this, education has an enormously important role to play.

Chapter 7
Practical Ethics in Achievement of Direct Reform

Abstract Activism is a growing force in changing society and animal activism is no exception. People like Henry Spira were influenced by earlier writings on animal rights and welfare, in particularly Singer's, and took up the cause. This chapter considers the cause of animal activism, as described in a variety of texts, and whether there are useful messages for eliciting change in future.

7.1 Introduction

Reform to improve the lives of animals can be achieved by direct action as distinct from legislative change. This is exemplified by the animal activist activities of Henry Spira in the USA from the mid-1970s to the late 1990s as described in his biography:

Ethics into Action: Henry Spira and the Animal Rights Movement by Peter Singer published in 1998

7.2 Contents of the Book

Henry Spira was born in Belgium in 1927. He spent the first half of his working life as a merchant seaman and subsequently as a secondary school teacher in the USA. His interest in human rights was reflected in his involvement while a seafarer with the Committee for National Maritime Union (NMU) Democracy that he continued for seven years after becoming a teacher (Singer 1998, pp. 1–44).

At the age of 45, Henry Spira had given little thought to animals (Singer 1998, p. 45). However, in 1973, he read the review essay by Peter Singer entitled "Animal Liberation" published in the *New York Review of Books* (Singer 1973, 1998, pp. xi, 47–48). The essay was submitted while Singer was still at Oxford, and it built on the book *Animals, Men, and Morals* edited by Stanley Gotlovitch, Roslind Gotlovitch, and John Harris published in Britain in 1971. This made Spira think seriously about

E.N. Eadie, *Understanding Animal Welfare*, Animal Welfare 13, DOI 10.1007/978-3-642-30577-1_7, animals as a group needing someone to act on their behalf. After reading the essay by Singer, it was recorded by Spira (Singer 1998, p. 49):

> Singer described a universe of more than 4 billion animals being killed each year in the USA alone. Their suffering is intense, widespread, expanding, systematic and socially sanctioned. And the victims are unable to organise in defence of their own interests. I felt that animal liberation was the logical extension of what my life was all about – identifying with the powerless and the vulnerable, the victims, dominated and oppressed.

Then, in 1974 Spira attended an evening adult education course conducted by Peter Singer at the University of New York to which Singer had moved. The course was based on the contents of the book *Animal Liberation* that was published the following year (Singer 1975). Spira was a person who saw knowledge as being more than just something for its own sake, and he had the attitude that "...if you see something that's wrong, you got to do something about it..." (Singer 1998, p. 51). During the course at New York University, Spira became a vegetarian, and at the end of the course, he initiated ongoing meetings at his apartment to do something practical about what he had learnt during the course. These were attended by several of the course participants including Peter Singer and Psychiatrist Leonard Rack, who provided valuable bio-medical expertise during the early campaigns for reform organised by Spira. It is observed by Singer (1998, pp. x–xi) in relation to the campaigns conducted by Spira:

> In twenty years, his unique campaigning methods have done more to reduce animal suffering than anything done in the previous fifty years by vastly larger organisations with millions of dollars at their disposal.

Indeed, the campaigns of Spira showed the enormous influence that can be exerted by an individual working largely alone but collaborating where appropriate with other individuals and groups. They showed what can be achieved by dedicated perseverance, clear thinking, and a realistic strategy. His campaigns exposed unethical practices involving animals and galvanised the power of public opinion. Moreover, the campaigns generally involved much more than just making people aware that what they were doing was ethically wrong. Indeed, Spira encountered opposition from scientists with vested career interests, company executives more concerned about profit and public image than they were about animal suffering, and government officials so complacent they had not even given a thought to the suffering of animals.

During the period of more than 20 years from 1975 to 1998, Henry Spira organised a range of successful campaigns directed primarily at eliminating or reducing the suffering of experimental animals and farm animals. In the case of laboratory animals (Singer 1998, pp. 45–139), the outcome of campaigns by Spira over 10 years from 1975 included:

1. The stopping of publicly funded research on the sexual aberrations of male cats whose brains had been subjected to surgical mutilation by Dr. Lester Aranson and Madeline Cooper at the prestigious American Museum of Natural History in New York
2. Cessation of the incongruous sponsorship by Amnesty International of medical experimentation on animals that included investigating effects of "torture" on pigs
3. The initiation by Revlon then Avon and Bristol-Myers of research to find alternatives to replace the cruel eye-blinding Draize test applied without

anaesthetic to the eyes of rabbits to test for eye damage to humans caused by cosmetics and other substances, accompanied by moves to get the government to accept alternatives in the certification of the consumer safety of such products
4. A consequential substantial reduction in the use of the Draize test following its replacement by a more refined tissue culture test
5. An enormous decrease in the use by Proctor & Gamble as well as other corporations of the flawed and cruel LD50 test for product safety that involved measuring the lethal dose of an ingested substance needed to kill half the animals in a group being tested, together with a great change in culture in regard to product testing that resulted in an enormous reduction in animal usage.

Indeed, by the end of 1989, eleven of the largest cosmetic firms in the USA had stopped product testing on animals entirely.

After 1985, Henry Spira turned his campaign activities to farm animals (Singer 1998, pp. 141–182), and his achievements for this category of animals included:

1. Exposing high-profile broiler chicken producer Perdue for false and misleading advertising about the living conditions of the chickens he reared for meat
2. Abolition in American slaughterhouses of the practice of shackling and hoisting cattle prior to Jewish and Moslem ritual slaughter and its replacement by an upright restraining device
3. Discontinuance of a requirement by the US Department of Agriculture that Mexican cattle imported into the USA be subjected to the cruel practice of hot iron face branding for the purpose of identification
4. Cancellation of a live duck and pheasant shoot planned to raise funds for the Helen Keller International organisation that was involved in programmes associated with blindness
5. Achieving progress towards getting McDonald's to adopt and enforce a policy that required the humane treatment of animals reared by suppliers to the restaurant chain
6. Establishment of the Centre for a Liveable Future to be devoted in a practical way to issues associated with sustainability and equity in a world heavily dependent on intensive farming, animal welfare abuse, and an unhealthy diet based on animal products.

These reforms in animal welfare achieved by the campaigns of Henry Spira involving both laboratory animals and farm animals were brought about by means other than reforming the law, and they showed how the discontinuance of unethical practices can result from direct action supported by the invoking of public pressure.

On the basis of the campaign experiences of Henry Spira, Peter Singer (1998, pp. 183–192) set out ten key points to assist future campaigners, whether working for animals or for other oppressed and exploited groups. In essence, these key points are:

1. Attempt to understand current public thinking and how it could be encouraged to develop. Importantly, maintain reality.
2. Choose targets for reform on the basis of vulnerability to public opinion, the degree of suffering, and the scope for change.

3. Set achievable goals and be prepared to bring about change step by step. It is not sufficient to merely raise awareness.
4. Use credible sources of information and documentation. Do not assume anything, do not deceive the media or public, and do not exaggerate or hype up an issue.
5. Remember that the world is not divided into those who are good and those who are bad, and try to empathise with others.
6. Aim to establish dialogue and work collaboratively to solve problems and attempt to present issues as solvable problems and try to suggest realistic alternatives.
7. Be prepared for confrontation if target is not responsive. If moderate approaches do not succeed, be ready to enhance public awareness campaign so as to put adversary on the defensive.
8. Avoid campaign bureaucracy and wasteful internal conflict.
9. Remember that legislation or legal action is not the only means of solving a problem.
10. Always assess critically whether what is being done will work in practice.

Spira did not work entirely alone, but he was the inspiration, the strategist, and the coordinator of campaigns that achieved important successes by eliminating the suffering millions of animals would otherwise have experienced (Singer 1998, p. 183).

It is appropriate to make some observations (Singer 1998, pp. 183–192) regarding the campaigning of Henry Spira that elaborate on the ten points listed above. In order to understand prevailing public thinking, Spira took every opportunity he could to speak with people outside the animal movement to avoid living in a world of unreality. Moreover, he selected targets for reform that took into account both the amount of animal suffering and the opportunity for change. He recognised it was necessary to do more than just raise awareness as important as this is and that the mere stating of an issue puts the target of a reform campaign on the defensive. It needs to be remembered also that no issue amounts to everything or nothing. Indeed, successful reform involves step-by-step incremental change. And the achievement of some success has a ripple effect that can lead to other and bigger successes as seen in the Revlon campaign that made research into alternatives reputable. At the outset of any campaign, there needs to be thorough preparation in terms of getting background information, and much such information is usually available in the public domain. It is preferable to use a publication of the target itself or a government document, but do not adopt material from animal activists' leaflets or from opponents of the target. At all times, a high level of credibility is essential.

When beginning any campaign, Spira approached the target with a polite letter inviting discussion about concerns he had. Moreover, in any attempt to get someone to do something differently, try to put oneself in the position of the other person and avoid a hostile approach. Indeed, try to empathise with the target and seek to suggest realistic solutions to any problem. Attempt to accentuate the positive side of anything. If a target is not responsive, confrontation becomes necessary. This can involve a public awareness campaign that might comprise newspaper articles, letters to the editor of a paper, picketing and demonstrations, or it might be possible to split adversaries, as Spira did in the case of the cat sexual aberration experimenters on the one hand and the museum employer and government provider funds on the other. In his operations, Spira avoided wasting time on committee meetings and money on bureaucracy by working largely alone, even though he did consult widely before making his own decisions. In general, Spira saw laws as maintaining the status quo and that they are changed merely to keep disturbances to a minimum. Indeed, only one of his campaigns was aimed at legislative change, namely the repeal of the New York State Metcalf-Hatch Act that allowed laboratories to source dogs and cats from shelters. Finally, every campaign needs to be directed towards a practical means of making a difference, and the question "Will it work?" always has to be asked.

7.3 Influence of the Book

It is difficult to distinguish the influence of the book *Ethics into Action: Henry Spira and the Animal Rights Movement* (Singer 1998) from the influence of the campaigns described in the book. Each demonstrates what can be achieved by direct action based on practical ethics to prevent animals from being subjected to suffering influenced by humans. Indeed, the successful campaigns of Henry Spira prevented millions of animals in a range of situations from experiencing enormous suffering they would otherwise have endured, and the lessons from the campaigns described in the book provide valuable inspiration and guidance for future campaigners to pursue direct reforms in animal welfare and to be more effective in their activist efforts.

Even though most of the reforms for animals achieved by Henry Spira resulted from direct action and not legislative change, reforms similar to those brought about by Spira have been enacted into the legislation of various countries. Indeed, it is likely that the direct reforms of Spira provided the original catalyst for such legal enactments. In South Australia, for instance, the *Regulations under the Prevention of Cruelty to Animals Act* 1985, as amended (No. 10 of 2000), forbid, though with exceptions, the use of practices on animals that corresponded to the Draize and LD50 tests without actually mentioning them by name, and in 1999, the face branding of cattle was banned in both Queensland and the Northern Territory of Australia (Stoner 2000, p. 11). Previously, Henry Spira had campaigned successfully in the USA against both the Draize and LD50 tests and the face branding of Mexican cattle entering the USA.

Subsequent to the campaign of Spira towards getting McDonald's to adopt and enforce a policy requiring the humane treatment of animals reared by suppliers to the company restaurants, McDonald's in the USA announced in the year 2000 that it would require its egg suppliers to provide a 50 % increase in the space allowance for each laying hen above the traditional space allowance per hen (Singer 2002, pp. x–xi), and in Britain, McDonald's later restricted its egg suppliers to those dealing only in free-range eggs (Sherman et al. 2005, p. 7). This was in response to consumer preference for humanely produced products, and McDonald's was given an RSPCA award for commitment to improving animal welfare. More recently, in Australia, the Hilton Hotel in Adelaide converted to using only free-range eggs in response to an RSPCA Australia *choose wisely* campaign (RSPCA South Australia 2009, p. 6; Innes 2009, p. 15). However, McDonald's in Australia is yet to convert to free-range eggs, but it is planning to do so. These examples illustrate the achievement of reform by direct action in response to public pressure and consumer preference and show that some progress has been made by McDonald's in relation to policy change requiring the more humane treatment of animals by its suppliers as initiated by Spira.

In the USA in 1980, John Mackey opened a grocery store in his garage in Austin, Texas, and this developed into what is now Whole Foods Market, with 300 stores worldwide, 54,000 employees, and annual sales of US$9 billion (Pacelle 2011, p. 319). Mackey ignored traditional business, thinking that shoppers leave their morals in the car park, and opened a business for shoppers he thought would consider and balance a range of factors including price, brand, quality, availability, and social responsibility in their purchasing habits rather than look merely at price regardless of other considerations. Indeed, many informed consumers do take into consideration factors such as whether foods are fresh and local, without pathogens or dangerous bacteria, and that animal-based foods are produced under humane conditions. This has resulted in an enormous increase in demand over recent years for food that is labelled *organic* or *humane*, and it shows that direct action by consumers through greater demand can result in conversion to animal-based food production involving greater humanity and reduced suffering. The rapid growth of Whole Foods Market is a demonstration of this, and it illustrates the power of direct action in achieving change to improve the lives of animals.

There have been various other types of change arising from direct action as distinct from legislative reform that resulted in improved animal welfare and less suffering, although they did not result directly from campaigns for reform. These include the creation of a computer model for the human heart by Oxford physiologist Denis Noble CBE FRS that led to a reduction in the number of animals used in experimentation (Noble 2004), the development of computer-based virtual teaching programmes for anatomical instruction by Alex Davies at Massey University in New Zealand (National Animal Ethics Advisory Committee 2004, p. 18), the use of immunocontraception in the population control of white-tailed deer in the USA to replace culling by shooting (The Advertiser 2006, p. 2), and the use of more humane scientifically based horse-training methods by Andrew McLean to train elephants in Nepal in a few days as an alternative to their traditional harsh training that took many weeks (Newby 2009). These examples illustrate the potential of the approach of Henry Spira in using direct action and appeal to ethical principles to reduce animal suffering.

A recent example of the power of direct action involving animal protection organisations and the media was the Australian Broadcasting Corporation *Four*

Corners programme "A Bloody Business" (Ferguson and Doyle 2011) shown on ABC Television on 30 May 2011. The programme exposed horrifying footage of the terrible cruelty to which exported Australian cattle were subjected in Indonesian slaughterhouses. It resulted in an enormous public outcry, and subsequently, a temporary ban was imposed by the Australian government on the export of live cattle to Indonesia. The evidence of cruelty to cattle at Indonesian abattoirs was videoed by Lyn White, a former Police Officer and now Campaign Director of Animals Australia, and it was analysed by Dr Bidda Jones, Chief Scientist of RSPCA Australia. The evidence was supported by *Four Corners*' own investigation. Because the Australian government had failed to act previously when shown disturbing evidence of cruelty to Australian animals exported to the Middle East, Animals Australia decided on this occasion to expose the evidence of cruelty to cattle exported to Indonesia directly to the Australian public through the *Four Corners* programme. There was a horrified and angry public reaction to the cruelty, and the government was quick to impose a ban on the export of cattle to Indonesia. The strength of people power in a democracy was evident, and clearly the Australian public cares deeply about major ethical issues involving animal welfare. Such a climate augers well for the success of campaigns to achieve reform in animal welfare by direct action as used by Henry Spira. Moreover, the approach of Animals Australia in its campaign against live animal exports from Australia bears some of the hallmarks of a Spira-type campaign.

7.4 Potential for the Book

In theory, at least the campaign approach of Henry Spira (Singer 1998) is applicable to any situation in which animals are subjected to human-inflicted suffering. This could include campaigning against the abuse of animals in agriculture, experimentation and testing, entertainment and sport, so-called pest control, or indeed in any situation in which suffering is imposed on animals. However, it is acknowledged by Singer (1998, pp. 184–186) that some targets are more amenable to a successful campaign than others and that reform often involves step-by-step incremental change. Indeed, some targets such as the powerful US agribusiness provide a difficult challenge, and such targets are well prepared and will resist any attempt to bring about change. It can help a campaign to be able to suggest realistic alternatives to what is being opposed, and if enough thought and lateral thinking are given to this, it is amazing what opportunities there are for change. This is illustrated by the examples of direct actions to improve the lives of animals given in the previous section. Moreover, public awareness of the needs of animals and opposition to their suffering, particularly in Western countries, is more favourable now than it was at the time Spira conducted his campaigns, and this provides an environment more open to change to reduce animal suffering. In countries without a democratic system of government, direct action through appeal to ethical principles might be the only means for achieving reform to create a better life for animals.

For the campaigning lessons of Henry Spira to be used, it is necessary for his strategies to be communicated to current animal activists. This can be effectively achieved by reading *Ethics into Action* (Singer 1998), and in this regard, it is good that the ten key points for activists proposed by Singer (1998, pp. 183–198) were reproduced in the book *In Defense of Animals: The Second Wave* (Singer 2006b, pp. 214–224) so they are drawn to the attention of a new crop of campaigners. Of course, it would be even more desirable for current animal activists to read the entire text of *Ethics into Action*. Fortunately, a good number of copies of the book are shown on WorldCat (2011) as being available in libraries in various countries, particularly in the USA where more than 500 libraries are listed as holding the book, and there will be copies of the book also in other libraries including those of animal protection organisations. Moreover, it is important that *Ethics into Action* continues to be available for purchase as it contains so many valuable lessons, and it is only by these being read and acted upon that the full potential of the book will be realised.

In her review of the book *Ethics into Action*, Glenys Oogjes, Executive Director of Animals Australia, observed (Oogjes 1998, p. 28):

> The last chapter 'Pushing the Peanut Forward' is the enduring lesson of Henry's remarkable life. The '10 ways to make a difference' are an inspirational instruction manual for success for animals (or other exploited and downtrodden beings). In my own view this book should be compulsory reading for anyone daring to call themselves an 'animal rights campaigner'. Then perhaps read it again when at first your goals are not achieved!
>
> Singer plays down his own considerable influence in Henry's awakening to the plight of animals and ever since, but make no mistake, this book is a collaborative work by our most influential animal rights philosopher and writer, and the most brilliant tactical animal campaigner of our time. To read it is to be inspired and to take a first step to intelligent and effective campaigning. Most of all the book will demonstrate to you that one person can make a difference, just as Henry has.

These words are as true today as they were when *Ethics into Action* was published in 1998. The important thing is for the book to be read and for the inspiration and lessons it offers to be put into practice to make a better world for animals.

7.5 Some Other Books

In *The Ethics of What We Eat*, Peter Singer and Jim Mason (Singer and Mason 2006) describe how animal welfare and the environment can be improved by ethical dietary decisions taken by people they describe as conscientious omnivores, who carefully choose the source of the food they eat on ethical criteria, and by vegans, who eat no animal products at all. The eating habits of these two groups are compared with the traditional American diet and the animal suffering involved, particularly in the case of intensively farmed animals, in producing the food for such a diet. Indeed, ethical decisions regarding what we eat can have direct and profound benefits for both animal welfare and environmental protection.

Also published in 2006 was the book *In Defense of Animals: The Second Wave* edited by Peter Singer, in which Part III of the book (Singer 2006b, pp. 157–224) deals with activists and their strategies. This part contains eight contributions covering various cases and issues ranging from the abolition in Austria of battery-farmed egg production in 2004 to take effect from 2009 (Balluch 2006, pp. 157–166) to reproduction of the ten points for activists (Spira and Singer 2006, pp. 214–224). In his review of *In Defense of Animals: The Second Wave*, John Webster (2006, pp. 192–193), in relation to the part of the book on activists and their strategies, observed a recurring theme that working through the power of public opinion and consumer behaviour was more expedient and effective than attempting to achieve reform through legislation. Webster agreed with this and in doing so provided a powerful endorsement for the use of direct action by a highly regarded academic veterinarian, who had served for many years on the Farm Animal Welfare Council (FAWC) in Britain and had observed closely the role of legislation in improving animal welfare.

The book *Making Kind Choices* by Ingrid Newkirk, President of People for the Ethical Treatment of Animals (PETA), published in 2005 (Newkirk 2005) is an easily read and practical guide to making direct choices in every aspect of human existence that result in a compassionate lifestyle with positive impacts on both the lives of animals and the environment. The book contains 78 short chapters of only a few pages each that convey compassionate messages relating to everyday living, and the chapters are arranged conveniently into ten parts of the book that deal with different aspects of life such as home and garden, fashion and beauty, and food and entertainment. According to Newkirk (2005, p. xxiii), the book shows a way to live life to its fullest and most meaningful by becoming a more caring person in many ways, big and small. Indeed, it is a valuable reference book to assist anyone wishing to adopt or spread compassion and kindness.

In his autobiography entitled *Committed: A Rabble-Rouser's Memoir* published in 2007, Dan Mathews, Vice President of People for the Ethical Treatment of Animals (PETA), provides some insight (Mathews 2007) into the campaign operations of PETA that attempt to achieve reform in animal protection by direct means rather than working for legislative change. Mathews believes that campaigning should be entertaining and fun in order to effectively bring animal protection issues to the people, particularly those people not influenced by traditional campaigning that merely exposes the facts. Moreover, PETA uses both sensational methods and public celebrities to draw attention to the plight of animals in various situations. In the book, Mathews (2007, pp. 119–141) describes in detail a successful campaign of PETA that resulted in ending the use of fur by a leading New York fashion designer Calvin Klein. Following a widely publicised sensational raid on his office in New York, Klein reluctantly agreed to view in the presence of Mathews a short video on the capture and killing of fur animals at the conclusion of which Klein declared he would stop using fur in his designs.

A book entitled *The Animal Activists Handbook: Maximising Our Positive Impact in Today's World* by Matt Ball and Bruce Friedrich published in 2009 is a valuable practical guide to direct campaigning (Ball and Friedrich 2009) written by two seasoned and successful animal activists, who held the positions of Executive Director of Vegan Outreach and Vice President of People for the Ethical Treatment of Animals (PETA), respectively. In Chaps. 1 and 2 of the book, Ball and Friedrich (2009, pp. 1–36) deal with general issues involving the choice of meaningful action and effective advocacy that apply to campaigning for any worthwhile cause. Then, in Chaps. 3 and 4, the authors address (Ball and Friedrich 2009, pp. 37–82) the specific matter of effective advocacy for animals and their own favourite ideas for changing the world. The latter include distributing leaflets and information stalls with a video such as *Meet Your Meat* produced by PETA, particularly on university campuses, but also at venues such as shopping centres where there is people traffic. Finally, in Chap. 5, Ball and Friedrich (2009, pp. 83–89) ask the question "Is Animal Liberation Possible?", and they conclude (Ball and Friedrich 2009, p. 88):

> Ultimately, *de facto* animal liberation will be achieved, but not with a bang. Change will not come by revolution, but through person-by-person outreach progressing hand-in-hand with advances in technology, leading slowly but inexorably to a new norm that, to most people, hardly seems different. But an unfathomable amount of suffering will be prevented.
>
> *It is up to us to make this happen.*

The book *The Animal Activist's Handbook* contains a wealth of intelligent and practical advice that can help achieve direct reform for animals and reduce the suffering inflicted on them by humans.

7.6 Overview

It is seen that direct action involving appeal to ethical principles and the power of public opinion can provide an effective means for achieving improvements in animal welfare by reduction in human-inflicted animal suffering. Such action is available in a wide range of situations involving the abuse of various types of animals including those used in agriculture, experimentation, leisure activities, pet breeding, and pest control. Activists wishing to improve the lives of animals need to consider any situation in which animals suffer at the hands of humans and examine the scope for direct action along the lines of the campaigns of Henry Spira (Singer 1998). Any such campaign for direct action will be enhanced by a courteous and cooperative approach to those involved in abuse, and any carefully thought through and viable alternatives to eliminate or reduce the animal suffering will help the campaign. In some parts of the world, particularly in Western countries, people are becoming increasingly aware of animal abuse and disapprove of it. This makes the climate for the achievement of improvements in animal welfare more favourable. However, even though several of the more obvious and blatant animal abuses have been successfully challenged as in Spira's campaigns, there remain some serious types of abuse covering enormous numbers of animals and a huge amount of animal suffering, as in the case of intensively farmed animals in the USA and other countries. These abuses frequently involve powerful vested interests such as the US agribusiness and its influential allies that resist change despite the animal suffering involved. Nevertheless, in such situations, animal activists have the moral imperative and growing public opinion on their side, and change for the better can be brought about, even if slowly, by perseverance and intelligent campaigning.

Already, there have been substantial changes as a result of direct action as seen in the demand for animal welfare-friendly foods and in other ways. Fortunately, animal protection organisations of various kinds are effective in identifying animal abuses that might otherwise go unnoticed, and they need to examine carefully the scope for direct action campaigns to improve the situation for animals involved. In order for them to benefit from the campaign experiences and successful approach of Henry Spira, there is much to be gained from reading his biography by Peter Singer (1998). Copies of the book are still available, and it is important that this continues to be the case. Even though all the campaigns of Spira were conducted in the USA, the principles he used can be applied anywhere in the world, and animal suffering needs to be addressed wherever it occurs. Indeed, in countries without a democratic system of government, direct action might provide the only means of improving the lives of animals. Of course, education has an important role to play in the effectiveness of direct action, whether by individuals in their own decision making or as part of an activist campaign to improve animal welfare. Education can enhance appreciation of animal sentience as well as recognition of the immorality of inflicting suffering on animals, and indeed the spreading of animal welfare education around the world is an important form of direct action to improve the lives of animals. There are many different ways in which direct action can be used to reduce animal suffering, and the successful campaigning of Henry Spira provides both inspiration and valuable lessons. We need to open our minds widely so we can envisage ways in which direct action can enhance the lives of animals by reducing their suffering. It is up to us to be imaginative, resourceful, and compassionate.

Chapter 8
Politics of Animal Protection

Abstract Community concern for animal welfare has the potential to lead to government legislation, principally at a national level. However, powerful interests in animal utilisation for profit have had significant influence on government policy, limiting attempts by the animal protection bodies to enhance the legislative framework. Indeed the animal protection bodies are often fragmented and have much less impact, even though it has been facilitated by the labour movement and increase of women in politics. In Britain there has been some government recognition of the widespread concern for animal welfare and rights in the formation of a Farm Animal Welfare Council, following the Brambell Committee that came about after the publication of Animal Machines by Ruth Harrison. This demonstrable government action on animal welfare in Britain has yet to be seen in the USA, but an increasing role for the European Union is emerging.

8.1 Introduction

The nature and extent to which the government in a democratic system enacts legislation or makes administrative decisions to improve the welfare of animals depends on political factors that influence its policy-making. An analysis of such factors is made in the book

Political Animals: Animal Protection Politics in Britain and the United States by Robert Garner published in 1998

8.2 Contents of the Book

In the book *Political Animals: Animal Protection Politics in Britain and the United States*, Dr Robert Garner of Leicester University, England, explores the nature of animal welfare policy-making in Britain and the USA (Garner 1998, pp. 1–2). In

E.N. Eadie, *Understanding Animal Welfare*, Animal Welfare 13,
DOI 10.1007/978-3-642-30577-1_8, © Springer-Verlag Berlin Heidelberg 2012

doing so his research seeks to address the questions of who makes decisions affecting the well-being of animals, who attempts to influence these decisions, why some decisions are taken rather than others, whether the decisions made are legitimate, and whether there has been a shift historically in the pattern of decision-making. Even though much had been written about animal protection issues by activists and journalists, Garner observed there were relatively few academic works on the subject and none by political scientists on public policy input. His book *Political Animals* (Garner 1998) was intended to redress this absence. Even though most animal-related policies are mentioned in the book, Garner (1998, pp. 10–11) concentrates particularly on the use of animals for food and in laboratory procedures because of the huge number of animals involved in these activities and the severity of suffering experienced by these types of animals.

According to Garner (1998, pp. 6–8), the adoption of *policy network analysis* is appropriate for studying animal protection decision-making as it is a policy area particularly subject to sectorisation in that a variety of institutional structures have evolved to deal with the different types of animal use, notably farm animals, laboratory animals, companion animals, and wildlife. An examination of these structures enables conclusions to be reached regarding the power relationships involved in animal protection decision-making, and it facilitates also an international analysis. In policy network analysis, it is assumed that public policy is not the product of government as a whole but rather derives from various networks within government, each with its own set of government and non-government players, behavioural norms, and power structures. Of importance here is the quality of access to sectors of the state by organisations having an interest in a particular sector's activities (Garner 1998, p. 7). This can vary from occasional contact with politicians and bureaucrats to regular formalised face-to-face decision-making. Indeed, it is the quality of access that largely determines the character of a particular policy sector.

In terms of structure, the book *Political Animals* (Garner 1998) comprises ten chapters, of which Chap. 1 (pp. 16–38) describes the existing sectorised administrative framework (including the legislation that has given rise to it) within which decision-making regarding the treatment of animals is made in Britain and the USA. In Chaps. 2–4 (pp. 39–108), the non-government actors involved are introduced, and in Chaps. 5 and 6 (pp. 109–138), an attempt is made to gauge the attitude of members of the British House of Commons and the US Congress towards animal protection issues. Then, in Chaps. 7–10 (pp. 139–228), the nature of animal protection decision-making is examined further in the case of specific sectors and national jurisdictions. Of these, Chaps. 7 and 8 (pp. 139–175) consider the politics of farm animal welfare in the USA and Britain, and Chaps. 9 and 10 (pp. 176–228) explore the politics of animal research in Britain and the USA. Finally, Garner makes conclusions (pp. 229–237) from the study regarding animal protection and pluralist politics.

Among the findings made by Garner (1998) in his study are:

1. In Britain and to a lesser extent in the USA, substantial legislative interest in animal issues has developed (Garner 1998, p. 38). Moreover, legislative and administrative actions have helped build sectorised institutional structures designed (superficially at least) to reduce human-inflicted animal suffering in a variety of ways.

2. In both Britain and the USA, those opposed to the introduction of further restrictions on their ability to exploit animals have developed formidable lobbying and public relations structures funded by wealthy non-government interests (Garner 1998, p. 67). Indeed, a significant number of politically active animal use organisations covering agriculture and research operate in both Britain and the USA as listed by Garner (1998, pp. 52–53).
3. A worthwhile strategy for the animal protection movement appeared to be that of pursuing reformist goals, whether or not these can be regarded as a means to ultimate animal rights objectives (Garner 1998, p. 92). Moreover, as animal welfare moves further towards the political mainstream, reformist objectives can be expected to involve increased concentration on conventional lobbying.
4. In contrast to wealthy and united (in the USA particularly) non-government organisations concerned with defending the right to continue exploiting animals, the animal protection movement appeared to have serious organisational, financial, and ideological problems (Garner 1998, p. 108). The best known non-government animal protection and wildlife conservation groups in Britain and the USA are listed by Garner (1998, p. 96). Of these, the organisations in Britain that were cited most in a survey as being influential were the Royal Society for the Prevention of Cruelty to Animals (RSPCA), Compassion in World Farming (CIWF), and the Royal Society for the Protection of Birds (RSPB), while the organisations in the USA cited most were the Humane Society of the United States (HSUS), People for the Ethical Treatment of Animals (PETA), and National Wildlife Federation (NWF).
5. In Britain animal welfare legislation and its administrative framework have developed largely independent of the parliamentary sphere (Garner 1998, pp. 109, 121). Nevertheless, while the direct legislative impact of parliament in Britain has been minimal, the legislative branch of government has contributed to the development of animal welfare as an issue. Moreover, the party and gender of Members of Parliament (MPs) are important explanatory variables in animal welfare issues, with labour and female MPs playing a greater role than others in animal issue advocacy in the House of Commons.
6. The US Congress plays a more substantial legislative function than the British House of Commons (Garner 1998, pp. 122, 137–138). However, Members of Congress in the USA have been less willing to raise animal welfare issues than their British counterparts, and the situation for animal advocacy in the USA was seen by Garner at the time of writing as somewhat depressing, particularly with a reduced number of Democrats in Congress.
7. On the matter of farm animal politics in the USA (Garner 1998, pp. 139–150), it is clear that farm animal welfare in that country is a victim of the policy-making process. It is a striking example of a policy arena dominated by interests with the most to lose by action to improve the welfare of farm animals. Indeed,

agribusiness interests are dominant in both the executive and legislative branches of government in the USA, and those seeking to promote farm animal welfare are correspondingly impotent (Garner 1998, pp. 139, 149–150). Nevertheless, there was a glimmer of hope in that there is some evidence of a move, even though slow, towards a greater concern among decision-makers in the USA regarding farm animal welfare. Moreover, the United States Department of Agriculture (USDA) had started to show greater interest in animal welfare issues, and in this, the successful campaigns of Henry Spira involving farm animals in the USA were an important catalyst. However, at the time of writing (Garner 1998), the dominance of agricultural interests within Congressional committees in the USA remained the key obstacle to reform. This was so even though agriculture committees had a relatively low status and positions on them were difficult to fill. Agribusiness lobbyists were aware of this situation and of increasing public concern for animal welfare. The response has been to focus on the need for industry to draw up its own codes of conduct in order to avoid legislative redress. However, agribusiness interests continued to possess impressive structural advantages, and their position was reinforced by Republican dominance of Congress. Indeed, the future for farm animal welfare in the USA remained unclear.

8. In Britain the relationship between agribusiness interests through the National Farmers Union (NFU) and the state is another good example of a policy community (Garner 1998, pp. 151–152). The privileged position of the NFU was enshrined in the *Agriculture Act 1947* that gave a statutory right to the NFU to be consulted by the Ministry of Agriculture, Fisheries and Food (MAFF) on policies that affected its interests. The glue for this relationship centred around the belief that agricultural intensification was desirable to provide reliable, cheap, and plentiful food, and factory farming was accepted as a central part of this ideology. Moreover, any groups such as consumers, environmentalists, and animal protectionists challenging this ideology were excluded from the centre of influence. Indeed, the Brambell Committee report of 1965 that arose from Ruth Harrison's scathing denunciation of factory farming was the first formal recognition by an official body that intensive animal agriculture raised animal welfare problems.

   The creation in Britain of the Farm Animal Welfare Council (FAWC) provided a forum in which both sides of the animal welfare debate meet, and the inclusion of industry representatives gave its recommendations greater legitimacy with government (Garner 1998, p. 165). Over the years FAWC members have included celebrated campaigners Clive Hollands and Ruth Harrison as well as sympathetic academics Marian Stamp Dawkins, John Webster, and Donald Broom. FAWC reports are available in the public domain, and relatively few of its recommendations are rejected, with a notable exception being its recommended ban on religious slaughter. Nevertheless, according to Garner (1998, pp. 174–175), there is significant evidence to suggest that British governments could have gone much further than they have in improving the welfare of farm animals, independent of whether or not the agricultural policy community is in a decline, and for more to be done in this regard, public pressure needs to be kept up and the consequences of intensive animal agriculture exposed. Indeed, public pressure has already resulted in weakening the sectorial nature of policy-making in the agricultural area as seen in the cases of the salmonella and BSE outbreaks as well as to some extent in the live export crisis that required a centralised government response. However, whether these issues were indicators of a permanent reshaping of the policy community in Britain or merely a blip on the landscape remained to be seen and was difficult to predict.

9. In relation to the politics of animal experimentation in Britain, the animal research lobby does have influence, but there are inadequacies in the legislative and administrative framework intended to protect animals used for research. Nevertheless, there has been a genuine government response to public concern, and the animal protection movement has played an important role in achieving legislative reform (Garner 1998, p. 176). An outcome of the *Animals (Scientific Procedures) Act 1986* was the establishment of a new statutory Animal Procedures Committee (APC) whose reports (unlike those of its predecessor) are available in the public domain, and this enables greater accountability by those administering the legislation (Garner 1998, pp. 182–192). In Britain, unlike the farm animal welfare policy network, the spread of influence within the policy arena involved with animal research is much more balanced. Indeed, Home Office Ministers dealing with animal research issues are much more likely to be influenced by the animal protection movement than are their farm animal colleagues in MAFF. Garner (1998, pp. 200–201) concluded that the animal protection movement came from nowhere to play a central role in the formulation, passage, and administration of the *Animals (Scientific Procedures) Act 1986*. The response of government to growing public concern about animal experimentation suggests that the policy community model is not applicable in the animal research policy arena. Even though the working of the new 1986 legislation has been far from perfect, it was clear that the most important positive development for research animals has been the facilitation of greater public accountability, particularly arising from the work of the APC. Indeed, public pressure possessed the potential in the future to result in further restrictions on animal research in Britain.
10. In the USA legislative and administrative controls on animal research are weaker than those in Britain (Garner 1998, pp. 202, 227–228). This is due in part to the influence of the animal research community, but in contrast to the public policy network surrounding farm animal welfare decision-making in the USA, its influence has been exercised in an increasingly complex pluralist policy network that has allowed the intentions of Congress and of public opinion to be thwarted. Nevertheless, in spite of this, Garner (1998, pp. 227–228) concluded that a positive outcome for animal advocates is that the issue of animal research has been politicised in the USA over the previous two decades or so. Indeed, animal researchers as never before have had to justify their work

to an increasingly sceptical public. Moreover, now, the issue of the ethics of animal use in research is being openly debated in the community, and Congress is conscious of this changed environment; it may well be difficult to counter pressure for improved implementation of existing law as well as further reforms relating to the treatment of laboratory animals in the USA.

11. It is concluded by Garner (1998, p. 235) that public pressure has led to reform in both Britain and the USA. On various occasions legislative change has occurred only after the materialisation of public concern, and frequently, this has been in response to a well-publicised event that demonstrated the need for reform. Moreover, public pressure is often generated and maintained by knowledge such as awareness of the public health and environmental implications of factory farming. The book *Political Animals* ends on an optimistic note (Garner 1998, p. 237):

> both the British and American political systems have shown themselves to be reasonably responsive to change. The task facing the animal protection movement in both countries is to mobilize, by whatever means, a large enough constituency to make further improvements in the lives of non-human animals a reality.

According to Mike Radford (1999, p. 303), the type, extent, and effectiveness of the protection given to animals by the state is a product of the political process, and an understanding of how the process works gives an insight into the influences that have shaped the existing regulatory framework and possibly suggests means for achieving further reform. In his review of the book *Political Animals*, Radford (1999, pp. 303–304) states that the book "provides a detailed and informative exploration of the character of animal welfare policy-making in Britain and the USA". However, Radford observes that the book does not sufficiently acknowledge the impact of the European Union (EU), whereas in reality British and EU policy-making, as in the case of agriculture, are intricately connected. Indeed, EU law places constraints on the governments of all Member States including Britain. Moreover, Radford notes that the book failed to make reference to the contribution of scientific research, yet this has been a crucial factor in developing the concept of animal welfare and our understanding of animals. Despite these two reservations, Radford (1999, p. 304) regards *Political Animals* as making an important contribution in a developing area of study and that the book should be of interest to all those concerned with the manner in which public policy in relation to animals is determined and implemented.

8.3 Influence of the Book

The book *Political Animals* analyses the nature of policy-making that lies behind legislative and administrative decision-making relating to the protection of animals in Britain and the USA. It shows that policy-making can be more than just a pluralist process in which power in a democracy is spread among a variety of competing interests with no particular interest dominating, and all interested parties getting some of their demands met according to the saliency of their views within the community as a whole (Garner 1998, pp. 5–6). Indeed, decision-making can be a sectoral process based on a variety of different networks, each with its own government and non-government participants, norms of behaviour, and power structure and with particular interests dominating in some policy arenas. The analysis of Garner elucidates the actors and structures involved and enables a greater understanding of how public policy decisions are made in different sectors of government. It demonstrates that decision-making to improve the lives of animals is not a one-sided process, exposes the nature of its adversaries, and possibly facilitates the adoption of procedures that can make the endeavours of activists to improve the welfare of animals more successful. The importance of transparency is seen by the effectiveness in improving animal protection that results from the reports of advisory bodies, notably the Farm Animal Welfare Council (FAWC) and the Animal Procedures Committee (APC) in Britain, being made available in the public domain.

Garner has shown that public pressure does result in reform, even if slow at times, and this can provide some encouragement to activists to continue with their campaigns. His study demonstrates that legislative reform most often follows enhanced public pressure resulting from wide public exposure of a situation involving animal abuse where reform is badly needed, and this indicates the desirability for activists to be vigilant in uncovering any such situation and ensuring it gets wide publicity. A good example of this is the recent television exposure on an ABC *Four Corners* programme (Ferguson and Doyle 2011) of the horrific cruelty to which exported Australian cattle were subjected in Indonesian slaughterhouses. Moreover, Garner emphasised the need for the animal protection movement in both Britain and the USA to mobilise more people in order to create a sufficiently large constituency to achieve further reforms. In order for the animal movement to be most effective, there should be unity of action among those involved, and in this regard, the recent cooperation between Animals Australia and RSPCA Australia in relation to the slaughter of cattle in Indonesia was an important factor.

In *Political Animals* (Garner 1998) it is seen that improvements in animal protection have been greater in Britain than they have been in the USA and that within the USA, they have been more evident in the research in the animal sector than in the agricultural sector. Thus, there can be greater achievements in improving animal welfare in one jurisdiction compared with another and in one sector compared with another in the same jurisdiction. This is so even though in each of the jurisdictions animals are legally regarded as property that is considered by Francione (1995) as the major factor inhibiting improved protection for animals.

The study by Garner (1998) has provided an important framework for understanding the politics of decision-making in relation to animal protection issues in Britain and the USA, and it should be useful for monitoring as well as participating in future developments. Moreover, it has established a basis for the making of similar studies in other jurisdictions, which is important as greater understanding of the nature of the political environment can result in more effective campaigning to create a better life for animals wherever they are located. The book *Political Animals* shows that the exertion of pressure in a democracy can bring about change to improve the lives of animals, and it suggests that the larger the animal movement constituency working for change, the more likely it will be achieved, even though in some situations change can be inhibited by powerful sectoral interests.

8.4 Potential for the Book

The book *Political Animals* (Garner 1998) is a valuable case study of the politics of animal protection in Britain and the USA, and it provides useful guidance for analysing public policy decision-making in the future not only in Britain and the USA but also in other jurisdictions. The study suggests a number of actions that have the potential for improving the protection of animals by political means. These include:

1. Understand the structure underlying public policy-making decisions related to animal protection in any particular jurisdiction.
2. Ascertain whether policy-making is sectorised, pluralist, or a combination of both.
3. Identify the main adversaries to change and how they operate.
4. Try to build up a sufficiently large animal movement constituency to pressure politicians to bring about change.
5. Adopt a unified approach to government in campaigning for reform, whatever one's ultimate aim, in order to be more effective.
6. Attempt to uncover cases of particularly appalling animal suffering, and get them wide publicity in the media.
7. Remain ever vigilant and maintain pressure on government.
8. Lobby for statutory provisions requiring advisory reports to government on animal issues to be made available in the public domain for reasons of transparency.
9. Try to encourage sympathetic politicians to become members of parliamentary committees, such as agriculture, that have influence in animal welfare issues.
10. Recognise that where direct action is not successful in achieving change, government legislative or administrative action might provide the only means available for bringing about reform.
11. Monitor, and if necessary pressure government, to ensure that any animal protection legislation is properly enforced.
12. Make sure that campaigns are well informed, and build on existing knowledge such as risks to human health or the environment.

Indeed, the book *Political Animals* embraces a number of important messages for activists relevant to improving the protection of animals by political means.

8.5 Some Other Books

In *The Political Animal: The Conquest of Speciesism*, psychologist Dr Richard Ryder (1998) examines the modern interest of humans in the welfare of non-human animals and in their rights or interests. He states that the question we have to face is "Are we prepared to continue to subjugate the principle of justice to our selfishness?" (Ryder 1998, pp. 1–3). The book comprises four chapters on history, ethics, animal welfare science, and politics. Of these Chap. 4 on politics (Ryder 1998, pp. 90–123) deals mainly with European animal welfare campaigns that had political objectives conducted from the 1970s to the late 1990s when the book was published. From his own personal experience of political campaigning from 1969 to 1986 for reform in animal experimentation law in Britain, Ryder (1998, pp. 97–100) observed that the ingredients required for success in campaigning included:

1. A high level of public concern
2. A well-argued case
3. Direct contact with ministers

At the end of the book, Ryder (1998, p. 126) concluded that animal welfare had become high on both academic and political agendas. Moreover, he asserted that all people had the potential for compassion and that the extent to which that compassion is manifested and nurtured depends on culture and experience. He regards suffering as the common enemy of human and non-human animals and that it provides the best focus for a new morality.

The book *The Animal Rights Debate: Abolition or Regulation?* (Francione and Garner 2010) comprises a debate between Gary Francione, Professor of Law at Rutgers University in the USA and a leading proponent of animal use abolition, and Robert Garner, Professor of Politics at Leicester University in England and a leading advocate of animal use regulation. Francione argues in favour of the animal rights approach in which he maintains there is no moral justification for humans using non-human animals and that such use should be abolished, whereas Garner argues in favour of the protectionist approach in which the permitted use of animals by humans is regulated to protect their morally significant interest in not being made to suffer incidental to their use (Francione and Garner 2010, pp. x–xi). The book contains three parts, namely a presentation and defence of the animal rights or abolitionist position by Francione (pp. 1–102), the case for the animal protectionist approach by Garner (pp. 103–174), and a discussion between Francione and Garner (pp. 175–269). In a review of the book, Mepham (2011, pp. 200–204) describes the engagement between Francione and Garner as enthralling, but one in which neither participant appeared able to concede ground to the other, so the divide between their respective positions remained. However, the book is useful in that it gives an authoritative account by leading exponents of the two disparate positions that have largely dominated the ethical underpinning of the animal movements in the USA and Britain respectively. It appears from the debate that much depends on what is meant by the word *progress*. Indeed, at the conclusion of the discussion between

Francione and Garner, it is stated by Garner in relation to progress (Francione and Garner 2010, p. 269):

> Our disagreement here crystallizes our respective positions. A great deal depends, of course, on how "progress" is defined and what time-scale is chosen. To finish on a note of consensus, we can both agree that animals endure unacceptable levels of suffering, that this suffering ought to be eliminated, and that encouraging a debate about the best means of achieving this goal is a priority.

To which Francione replied: "Agreed". Clearly, both Francione and Garner concur that much needs to be done to improve the lives of animals that suffer at the hands of humans and that the best way for achieving this needs to be found. According to Garner (Francione and Garner 2010, p. 105), it is important to distinguish between ethical prescriptions and what is achievable politically and strategically. Indeed, it is argued by Garner (Francione and Garner 2010, p. 168) that even though there is a strong philosophical case for granting at least some rights to animals, it is not a realistic possibility for animal rights goals to be achieved at present. Moreover, the reality is that most national animal rights organisations advocating the abolition of animal use do in fact adopt strategies supporting animal protectionist reforms as a means towards achieving their ultimate abolitionist goal (Garner 1998, pp. 83, 92).

An enlightening account by an insider of the quest to improve animal protection in the USA is contained in the book *The Bond: Our Kinship with Animals, Our Call to Defend Them* by Wayne Pacelle (2011). For the previous 17 years, Pacelle had worked with the Humane Society of the United States (HSUS), the largest animal protection organisation in the USA, and for the last 7 years, he was President and Chief Executive of HSUS. Pacelle is a graduate of Yale University and has taken a special interest in law reform in which he has played a leading role as strategist in the enactment of animal protection laws by direct action of the electorate in several states of the USA that have outlawed factory farming practices, cockfighting, bear-baiting, and various other inhumane activities.

The book *The Bond* (Pacelle 2011) contains chapters on animal slaughter and factory farming (Chap. 3, pp. 91–134) and on the defenders of animal cruelty (Chap. 7, pp. 269–311). These chapters include a discussion of the successful citizen-initiated ballot for Proposition 2 in California (a major factory farming state) that in 2008 outlawed from 2015 the extreme confinement of veal calves in crates, laying hens in battery cages, and pregnant sows in gestation crates (Francione and Garner 2010, pp. 142–143) and of the parts played by the United States Department of Agriculture (USDA) and the American Veterinary Medical Association (AVMA) in supporting the agriculture lobby in the USA. Even though *The Bond* is extremely broad in its coverage of issues relating to the relationships between humans and animals and of the need for changed human attitudes to provide better protection for animals, the book does give a fascinating and valuable insight into the politics of animal protection in the USA as observed by Wayne Pacelle during his efforts with the HSUS to provide a better life for animals.

8.6 Overview

In a democratic system of government, public policy decision-making in relation to animal protection can be pluralistic, sectoral, or a combination of both. Pluralistic decision-making involves the whole of government, and public policy outcomes are generally a compromise between competing interests so that change is often a slow incremental process with much depending on the influence and arguments of the disparate parties affected by decision-making outcomes. Unless the animal protection movement develops a large constituency and a dominant voice behind such decision-making, no rapid and dramatic improvements in the protection afforded to animals can be expected, quite apart from any ethical arguments for more rapid progress, and this is particularly so as strong vested interests opposing improvements in animal welfare are likely to remain. Indeed, any changes for the better in animal welfare are likely to involve compromise and incremental change. Nevertheless, it is important for the animal movement to develop a strong and unified constituency to achieve whatever improvements it can, and there are signs that electorates, in Western countries at least, are becoming more concerned about animal welfare issues and politicians are aware of this.

In the case of sectoral decision-making much depends on existing structures and the influence of government and non-government players involved in the process. Historically, powerful vested sectoral interests as in the case of agriculture in the USA have dominated decision-making and thwarted improvements in animal welfare. Only if the animal movement is able to develop an overriding influence in any sectorised decision-making, a situation that seems unlikely, could rapid improvements in animal protection be achieved in a sectorised system. Thus, whether public policy decision-making is pluralistic or sectorised, it is unlikely that there will be any sudden and dramatic improvement in animal protection as a result of public policy decision-making in a democratic system. Nevertheless, governments have responded at times to address particularly serious cases of animal suffering, especially where there has been wide publicity and public outcry. Generally, however, animal welfare issues have to compete for the attention of government with a great many other diverse and complex issues that occupy the time of politicians, particularly ministers.

Despite the weaknesses in a democratic system for achieving legislative change to improve animal welfare, there have been gradual incremental improvements in the legal protection of animals, particularly in Europe, and these are likely to continue. In order to achieve the best possible outcome, the animal movement needs to be vigilant, unified, and outspoken, and it has to develop as large a constituency as possible to maximise its influence on government. Moreover, animal suffering is universal, and animals are abused all over the world, so the animal movement needs to be strong globally and exert whatever influence it can using political means in democratic systems wherever they exist anywhere in the world. In this regard, the large international animal protection organisations play an important role, but their efforts have to be supported by strong domestic campaigning. Furthermore, any positive lessons learnt for improving animal protection in any particular jurisdiction can be used as a guide to achieve change in others, so a watching brief should be kept on positive developments everywhere. In any event, an analysis of the politics of animal protection as made by Garner (1998) for Britain and the USA should enable a greater understanding of the environment in which public policy decisions relating to animal protection are made in any specific jurisdiction, and this can assist in making political campaigning more effective.

The achievement of improvements in animal welfare by political action should not be regarded as the only means of improving the lives of animals but should be seen as complementary to the use of direct action discussed in the previous chapter. Indeed, every possible means for improving animal welfare should be considered and used. This includes the adoption of vegetarian or vegan eating habits that not only eliminate the animal suffering that would otherwise have occurred but contribute also to improved human health, environmental protection, and global equity. However, not everyone is going to become vegetarian or vegan; indeed, meat eating is increasing globally, and not everyone will make purchasing decisions based on compassion. Moreover, there will be powerful vested interests resisting any change, no matter how strong the moral arguments are for this to happen. Thus, political action resulting in legislation, regulation, and administrative decisions might be the only means in some cases of achieving improved animal protection.

In some states of the USA and in other jurisdictions, provision exists for citizen-initiated referenda that provide the opportunity for the electorate to propose and vote on particular issues, as was the case with Proposition 2 in California in 2008 related to factory farming in that state. This resulted in a successful vote to ban by 2015 the extreme confinement of intensively farmed animals including calves, pigs, and hens (Francione and Garner 2010, pp. 142–143; Pacelle 2011, pp. 119–134). Essentially, Proposition 2 will mandate the basic freedoms relating to space requirements for animals recommended by the Brambell Committee in Britain in 1965. The outcome of Proposition 2 should be improved animal welfare resulting from a direct vote that reflected the will of the electorate without the disadvantages inherent in the traditional political process. However, not all states in the USA have citizen-initiated referenda, and a vote for change in one state only can provide an incentive to move activities to another state. Nevertheless, the vote in California is an important step that puts intensive farming firmly on the political agenda, and it will probably act as a catalyst for further improvements in the treatment of farm animals in California and elsewhere. Moreover, citizen-initiated referenda could be introduced in more states of the USA and in other countries, and they could prove to be a useful political means of improving animal protection, particularly as there is growing awareness of and concern for animal welfare issues in the community. In any event, it is desirable to adopt every means possible, whether political or direct, that can create a better life for animals.

Chapter 9
Legislative Enactment for Animal Protection

Abstract The development of animal welfare law in Britain can be traced back to before the publication of Animal Machines and eventual formation of the Farm Animal Welfare Council. The first vestiges appeared about two hundred years ago. Although in Britain a dominant role in promoting welfare through legislation and other methods of control is evident for the RSPCA, the emergence of international bodies, such the EU, the Council of Europe, the World Animal Health Organisation (OIE), and the WTO, is rapidly subsuming such national interests in relation to responsibility to animals. The advent of legislation in animal welfare was necessitated by changing public ethics, in particular relation to the minimisation of cruelty and restricting unnecessary suffering.

9.1 Introduction

Animal welfare can be improved by the enactment and enforcement of laws that provide for the protection of animals from suffering inflicted by humans. An authoritative account of the evolution and role of such legal protection for animals in Britain is contained in:

Animal Welfare Law in Britain: Regulation and Responsibility by Mike Radford published in 2001

9.2 Contents of the Book

In the Preface of *Animal Welfare Law in Britain: Regulation and Responsibility*, it is observed by Dr. Mike Radford (2001, p. ix) of Aberdeen University that our growing understanding of the profound implications of the view that humans and other animal species evolved in the same way and had shared progenitors is possibly the most important factor in appreciating animal capabilities and needs.

E.N. Eadie, *Understanding Animal Welfare*, Animal Welfare 13,
DOI 10.1007/978-3-642-30577-1_9,

It has contributed to reassessment of our responsibility towards animals and provided impetus for the development of legal regulation. Initially, such regulation was to provide protection to animals from cruelty, and more recently, it has been able to achieve higher animal welfare standards that enable improved quality of life for animals. A body of law designed to create protection for animals has developed in Britain over a period of almost two centuries, and it will continue to evolve. The book *Animal Welfare Law in Britain* is intended by Radford (2001, pp. x–xi) to provide a progress report giving a broad overview of animal protection law in Britain and importantly to make a contribution towards understanding why legal regulation is necessary, how far the law has progressed, and what are its weaknesses.

In terms of structure, the book *Animal Welfare Law in Britain: Regulation and Responsibility* (Radford 2001) comprises six parts. These are:

Part A: Introduction (pp. 1–11)
Part B: Historical Development (pp. 13–95)
Part C: The Legal and Political Context (pp. 97–192)
Part D: Cruelty (pp. 193–258)
Part E: Welfare (pp. 259–341)
Part F: Effectiveness (pp. 343–405)

In Part A Radford (2001, pp. 1–11) outlines how in around 200 years the situation relating to animal protection in Britain has changed from one of common disregard for animal suffering and until 1822 the absence of any legal protection for animals to one of state intervention to protect animals culminating in almost three and a half thousand legal provisions relating to animals by 2001. The focus in the book (Radford 2001, pp. 9–10) is on human responsibilities, not animals' rights, as Radford believes the protection of animals in the foreseeable future can be best served by an incremental approach that includes imposing detailed and binding positive duties on those responsible for them. This assumption is consistent with the regulatory approach of Robert Garner (1998) but differs from the abolitionist approach of Gary Francione (1995). In Part B of the book, Radford (2001, pp. 13–95) deals with the development historically of animal protection legislation in Britain. This includes a consideration of the challenges to the anthropocentric tradition regarding animals and the need for protective legislation, achievement of the first such legislation in 1822 and establishment of the Society for the Prevention of Cruelty to Animals (SPCA) in 1824, and subsequent extensions to legislative protection that included the *Protection of Animals Act 1911*. In Part C (Radford 2001, pp. 97–192) consideration is given to the contemporary legal and political context. This includes an examination of the continuing need for legal regulation and consideration of the current legal framework that embraces the World Trade Organisation (WTO), the Council of Europe, and the European Union. Moreover, the various domestic sources of regulation are described, and the important relationship between law and politics is discussed, in particular, the means for achieving legislative change. Parts D and E (Radford 2001, pp. 193–258, 259–341) deal with the substantive law on cruelty and welfare respectively, although this is a somewhat artificial distinction as observed by Radford (2001, p. 6) in that the welfare of an animal subjected to cruelty is adversely affected. Nevertheless, there are important differences in that cruelty is a *legal* concept and animal welfare is a *scientific* discipline. Moreover, cruelty is generally *negative* in character, whereas animal welfare is somewhat more *positive*. In Part F Radford (2001, pp. 343–405) covers the matter of enforcement of legislation that is essential to make it effective. Finally, a number of issues discussed previously are brought together and suggestions made for possible future reforms. In addition, the book contains a table of cases and tables of legislation (Radford 2001, pp. xvi–lxvi) covering various jurisdictions.

The conclusions reached by Mike Radford in *Animal Welfare Law in Britain* (Radford 2001) include:

1. The introduction of early animal protection legislation in Britain during the first third of the nineteenth century was based on a number of factors including improved understanding of animal physiology, reassessment of the place of humans in the world, increased influence of urban middle-class values, endorsement by the upper echelons of society, the actions of individual campaigners in furthering the cause of animal protection, and a responsive political and legislative system (Radford 2001, p. 59). Indeed, the influence of the old rural community and its values was replaced by that of a new urbanised society governed from London, and despite much opposition, a centralised national morality overrode local traditions including established property rights so as to incorporate into law the protection of animals against cruelty.
2. During the period of less than a hundred years prior to the First World War, a complex and extensive legislative regime had been built up in Britain to provide protection to a range of animal types against cruelty (Radford 2001, pp. 93–95). These included domestic and captive mammals, birds, reptiles, and fish, and the perpetrators of cruelty as well as others responsible for the animals were subject to prosecution with focus being put on the effect of cruelty on the animal rather than the characteristics of the perpetrator. Liability could arise from acts of omission as well as of commission, and the law covered mental as well as physical suffering. Indeed, animals of various types in a range of circumstances were provided some protection, and these developments amounted to a "revolution" in the attitude of the state to other species. Moreover, in the courts, some judges endorsed legal intervention to prevent cruelty to animals on the grounds it would improve human character, morality, and education, while other judges recognised that humans owed a direct moral duty to minimise animal suffering (Radford 2001, p. 94).

   Study of the social and legal history of humans' changing attitudes towards animals is not only fascinating in its own right, but as observed by Radford (2001, pp. 94–95), some knowledge of it is essential for a proper understanding of contemporary law. For various reasons the form of animal protection legislation in Britain at the start of the twenty-first century had its roots in legal developments during the nineteenth century. These include the ad hoc nature of many provisions enacted in response to particular situations, the importance of lobbying to achieve change, the responsibility for criminal law enforcement being shared between public authorities and charitable organisations, the existence of separate legislation and varying degrees of protection to animals applying in the constituent parts of the UK, and the spread of responsibility for animal protection and its administration between a number of different government departments. Indeed, knowledge of the past is necessary to comprehend the present.

3. Radford (2001, p. 102) asserts that legal regulation of the manner in which animals are treated by humans continues to be essential in order to counteract the otherwise unconstrained property rights of the owner of an animal under common law. However, Radford (2001, p. 129) regards as overstated the importance given by Gary Francione (1995) for the need to alter the legal status of animals as property, and he observes that some of the claims by Francione are not supported by experience. Indeed, Radford (2001, p. 129) sees the issue of the legal status of animals as something of a distraction and that it is more important to achieve a level of legal regulation for animals that recognises their needs and capacities, confers a clear and precise duty on those responsible for them so the intended standards can be properly applied, and ensures that the strategy is supported by adequate enforcement measures. It is regarded as important by Radford (2001, p. 118) to be able to demonstrate an ethical and scientific basis for legislative intervention designed to create better protection for animals. Indeed, ethics and science should provide the *justification* and law the *means* for improving animal protection. Nevertheless, it is recognised by Radford (2001, pp. 126–128) that other means also can be used to complement legal regulation, although they are not a substitute for it. In considering the application of any legislative provision, it is necessary to establish by reference to the statute itself, and any secondary legislation made under its authority, the animals to which the legislation applies (Radford 2001, pp. 124–126). Moreover, much depends on the interpretation given by the courts to terms such as *unnecessary* that can result in differing standards depending on whether a particular animal is, for instance, a farm animal or a companion animal. Thus, the degree and nature of protection given to an animal under the law depend very much on the situation in which the animal finds itself rather than on its individual needs (Radford 2001, pp. 128–129), and this regulatory relativism is something that needs to be addressed by policy makers. However, law reform is a protracted process (Radford 2001, p. 127), and it is influenced by political expediency more than it is by principle.
4. The practical operation of animal protection legislation in Britain requires an appreciation of the nature, source, and status of the different types of law that apply as well as of the relationship between them as observed by Radford (2001, p. 165), and he provides an insight into this (Radford 2001, pp. 131–165). Traditionally, the United Kingdom (UK) Parliament enjoyed legislative supremacy, but in more recent decades, this autonomy has been severely diminished by the *European Communities Act 1972* under which the

UK ceded legal authority to European Community institutions on matters falling within the competency of the Community and by the *Scotland Act 1998* and the *Government of Wales Act 1998* under which the UK devolved both legislative power and executive accountability to both Scotland and Wales, particularly to the former (Radford 2001, pp. 141–151, 159–160, 165–166). Consequently, it is now necessary to lobby beyond Westminster to achieve reforms in animal protection legislation. Moreover, the trade liberalisation ideology of both the World Trade Organisation (WTO) and the European Community (Radford 2001, pp. 132–137, 165–166) has important adverse implications for animals used commercially in that it threatens existing gains and future progress relating to improvements in welfare for such animals, particularly farmed animals. Thus, the traditional supremacy and independence of the British legal system appear in practice to have been superceded, and this has important implications for the operation of representative democracy and the matter of animal protection in particular.

5. It is observed by Radford (2001, p. 167) that legislation is the product of the political process. Even though scientific advice is generally essential and often paramount in decision-making, there are times when there is no scientific consensus. Moreover, ethical, economic, political, and health issues may need to be taken into account also and judgement exercised by politicians as decision-makers. Indeed, both the substance and form of legislation are the product of circumstance as much as of principle. In the absence of scientific knowledge or certainty, there is strong moral argument (Radford 2001, p. 192) for application of the precautionary principle so that animals are given the benefit of the doubt to protect them from suffering. The impetus for legislative reform can come from a range of different sources (Radford 2001, pp. 175–177), and these include, among others, pressure groups, the government itself, and standing committees created specifically to advise ministers.

   With the accession of the UK to the European Community in 1972, the attention of lobbyists, previously concentrated on Westminster and Whitehall, had to turn to Brussels and Strasbourg (Radford 2001, pp. 183–190), and their activities make a substantial contribution to the working of the Community. Indeed, pressure groups that seek to influence public support for improved animal protection perform an important democratic function within the Community that includes helping *inform* the legislative process, *publicising* what is happening at Community level, encouraging *participation* in the legislative process, and acting as *agitators* by monitoring the actions and outcomes of Community institution activity. The Eurogroup for Animal Welfare (Eurogroup) plays a central role in relation to animal protection matters within the Community (Radford 2001, pp. 184–185), and it provides the Secretariat for the European Parliament's all-party group on animal welfare known as the Intergroup on the Welfare and Conservation of Animals. The animal welfare lobby within the Community has conducted sophisticated and successful campaigns to influence politicians. Moreover, in relation to putting animals into politics, public opinion has been well mobilised so that politicians as well as government officials in both

Britain and the European Community now recognise that animal welfare is an important issue of public policy.

6. The animal protection law prevailing in Great Britain in 2001 was in England and Wales based largely on the *Protection of Animals Act 1911* and in Scotland on the almost identical *Protection of Animals (Scotland) Act 1912* (Radford 2001, pp. 195–219). These statutes set out the legal meaning of cruelty, and the 1911 Act primarily consolidates the existing law into a single statute. The cruelty offences were central to the statutory protection of animals in Britain. They were wide in scope and largely effective but reflected the nineteenth century origins of the 1911 and 1912 Acts. It was suggested by Radford (2001, p. 219) that some minor amendments could have a significant effect in improving the legislation. In particular, the *prospective element* of cruelty contained in the offence of abandonment could be extended to all cruelty offences based on causing unnecessary suffering by adding the words *or likely to*, although the government had recently rejected such a proposal.
7. The longevity of animal protection law in Britain is a reflection of its effectiveness, and the legislation generally worked well (Radford 2001, pp. 221–240). When the suffering of an animal has been established, there is rarely an excuse for it. If not inflicted deliberately, it is usually possible to prove the defendant knew or must have known about the condition of the animal. In most offences of cruelty, an objective test is applied in which the behaviour of a defendant is judged by the standards of a reasonable person in the same situation, and the offences extend beyond deliberate infliction of suffering to include negligence, neglect, or disregard of an animal's needs or interests. This test has the advantages of retaining a degree of judicial discretion while preventing members of the court from being influenced entirely by their own personal values and attitudes and particularly importantly in providing better protection for animals by disregarding the motives, standards, and circumstances of the defendant. The latter factors are not taken into account in determining guilt but are relevant in deciding whether to prosecute and the severity of the punishment in the event of guilt. Moreover, the 1911 and 1912 statutes not only penalise the deliberate and wanton abuse of animals but promote responsible attitudes towards the care and treatment of animals as judged by the standards of the reasonably caring and humane person. Nevertheless, the concept of *unnecessary suffering* had to be considered in the context of the animal protection legislation.
8. The *Protection of Animals Act 1911* was a consolidation statute so that the case of *Ford v Wiley* decided in 1889 remained good law. It is legal authority that *necessity* provides a good reason to cause an animal to suffer. Establishing such necessity involves a balancing exercise in which *all* relevant considerations must be taken into account. A summary of the factors to be considered in deciding whether suffering is unnecessary is given by Radford (2001, pp. 257–258). The concept of unnecessary suffering has been developed by the courts and adopted widely by legislation. It has the important merits of being applicable to a multitude of different situations and can be constantly reinterpreted by the courts as knowledge about animal suffering develops and

social attitudes regarding the proper treatment of animals change. This obviates the need to regularly amend and update legislation. However, even though the prohibition on causing unnecessary suffering has contributed in a major way to improving the treatment of animals by humans, it is essentially negative in character in that it focuses on what should not be done to an animal. The concept remains central to providing legal protection to animals, but in recent years, greater focus has been placed on the welfare of animals and the imposition of positive duties to improve their welfare. Indeed, as observed by John Webster (Radford 2001, p. 258), "the care of animals involves more than just the absence of cruelty".

9. Radford (2001, pp. 261–287) provides a broad overview of the developing concept of animal welfare and identifies some of the issues being debated by those involved in the field. The issue of animal welfare had grown substantially in importance over the previous 30 years or so, and by the end of the previous century, it had become the major factor influencing public policy relating to animal treatment, particularly in the matter of legislative reform. Even though there is an overlap between animal cruelty and welfare, cruelty is a long-established *legal* term, whereas welfare is essentially a *scientific* concept that had its origins in the Brambell Report and is currently epitomised in what have developed into the Five Freedoms for animals. The fact that some animals at least are capable of suffering gives their welfare a moral dimension. Even though the minimisation of suffering is the principal aim of those concerned with animal welfare, the absence of suffering is not entirely definitive of good welfare. Legislation for animal welfare consists primarily of measures that seek to regulate activities involving animals by prescribing means for their treatment and care.
10. The extent of public control over animals and those responsible for them is discussed by Radford (2001, pp. 289–316), and he shows there is wide variation in the degree of sophistication among the various schemes adopted. This ranges from a highly detailed and complex system regulating the use of animals in scientific procedures to the absence of any formal scheme of licensing or certification for circuses, markets, and farming. Further, many of the older licensing schemes, particularly those relating to companion animals, were in urgent need of update. In the case of professional proficiency, veterinarians are regulated by the Royal College of Veterinary Surgeons (RCVS) under the auspices of statute, but the RCVS does not have the necessary powers to routinely monitor members of the profession and hold them to account. Moreover, the underlying legislation implies that a generalist veterinary qualification adequately equips a member to undertake any aspect of veterinary work, whereas there have been extensive developments in veterinary and welfare science since the mid-1960s. In contrast, the most significant development in regulation of competency is that related to animal transporters, and this could provide a model for other activities such as farming and dog breeding where appropriate or adequate regulation did not exist. Radford (2001, p. 316) concludes that the practical and symbolic significance

of licensing, certification, and registration relating to both undertakings and personnel should not be underestimated as they restrict the traditional right of an animal's owner to do as he or she sees fit. Indeed, they have the potential to provide a substantial buffer against the consequences of the common law status of animals as property.

11. Legal regulation of the quality of care for animals is discussed by Radford (2001, pp. 317–341), and it is seen that in Britain, substantial and increasingly sophisticated legislation prescribing standards relating to the care and treatment of animals was enacted. This legislation owes much to scientific research on the needs of animals and the benefit to them of high welfare standards. However, some older legislation lagged behind appropriate welfare requirements, particularly that relating to companion animals, and there was urgent need to enhance the standards required for pets. This falls outside the jurisdiction of EC law that covers the welfare of animals in other categories. Indeed, in Britain, the Protection of Animals Acts has been relied upon in relation to the treatment of companion animals, but these are concerned largely with cruelty. Moreover, even though there have been many important developments in welfare legislation, such legislation continued to sanction practices many people regard as unacceptable, particularly in commercial situations where animal interests are compromised for human benefit. Nevertheless, the introduction and ongoing development of animal welfare legislation in Britain provide an important extension to the protection provided to animals.
12. For legislation designed to protect animals to be *effective*, it must be adequately *enforced* (Radford 2001, pp. 345–395). This involves much more than just bringing offenders before the court, which should be regarded as a last resort. Indeed, proper enforcement involves ensuring (Radford 2001, p. 345):

    (a) That any regulatory schemes, whether prescribed by the European Community, the national Parliament, the Scottish Parliament, or the Welsh Assembly, are properly implemented
    (b) That those to whom the legislation applies know the nature and extent of their legal responsibilities
    (c) That animals are treated according to the standards required by legislation
    (d) That advice and guidance is given as to how animal care can be improved to satisfy the standards required
    (e) That, if necessary, appropriate authorities exercise their powers to ensure animal welfare does not continue to be compromised

    Moreover, legislation should be certain in its object, application, and consequences (Radford 2001, pp. 391–392); any measure prescribed should be appropriate to its subject matter and purpose; and responsibility should extend to *all* those whose conduct has a bearing on the welfare and treatment of the relevant animals. Those responsible for enforcement should have the necessary authority and powers to ensure compliance with the law, and they should have the relevant training, knowledge, and independence. Further, the sanctions available should include measures such as confiscation and disqualification orders to protect animals. Administrative measures such as improvement notices and licensing powers are also important. Finally, statutory enforcement agencies should be adequately resourced to properly execute their responsibilities. Radford (2001, p. 392) asserts that, even though recent enactments had the potential to improve enforcement, the present regulatory framework was unsatisfactory in terms of the criteria outlined above. Indeed, without the activities of the Royal Society for the Prevention of Cruelty to Animals (RSPCA) and the Scottish Society for the Prevention of Cruelty to Animals (SSPCA), enforcement would have been totally inadequate. These charitable organisations operate without public funding or any special legal power. However, the major contribution these organisations make to animal protection should not be allowed to absolve statutory agencies from their responsibility, and such agencies should be publicly accountable with enforcement open and transparent. Moreover, the publication of appropriate annual statistics by all public agencies involved in welfare enforcement would assist in producing greater accountability. With the availability of such information, the public could judge whether statutory organisations responsible for enforcement were performing their duties adequately, and the community, politicians, and policy makers could make an informed judgement as to whether further legislative reform was necessary.

13. Radford (2001, pp. 395–397) advocated that the Protection of Animals Acts that cover domestic and captive animals should be extended to create a duty of care on the owners and keepers of such animals similar to that recently imposed on the owners and keepers of farm animals under which they are required to take all reasonable steps to ensure the welfare of the animals under their care. Moreover, other suggestions are made by Radford (2001, pp. 392–402) for reforming domestic law to improve the legal protection provided to animals. However, it is recognised by Radford (2001, pp. 402–405) that the time had passed when animal protection law was exclusively determined by the UK Parliament and that European Community institutions, the World Trade Organisation, the Scottish Parliament, and the Welsh Assembly together have a profound effect on the law and impose significant restrictions on the legal power of and unilateral action by Westminster. Finally, Radford (2001, p. 405) concludes that progress in securing adequate protection for animals may seem slow and frustrating, but he recognises that evidence, time, and persuasion are required to move forward. However, much has been achieved since the pioneers of legal protection for animals embarked on their novel quest, although much remains to be done. Indeed, the evolution of society historically has been marked by increasing concern for the welfare of animals, particularly with enhanced human prosperity and the industrialisation of animal production, and it appears reasonable to anticipate that this concern would continue to develop so that conditions tolerated today may be regarded as intolerable in the future as observed in 1965 by the Brambell Committee and quoted by Radford (2001, p. 405).

9.3 Influence of the Book

The book *Animal Welfare Law in Britain: Regulation and Responsibility* (Radford 2001) has been able to exert influence in a variety of ways. These include:

1. The book provides a scholarly, comprehensive, and detailed account of the evolution and contexts of animal protection law in Britain.
2. It is written in such a way that a complex and difficult subject is made interesting and easy to understand.
3. It shows how animal protection law in Britain has developed from a complete absence of protection at the beginning of the nineteenth century to the existence of extensive legal provisions to prevent cruelty to animals and more recently the enactment of welfare standards for the proper care and treatment of animals.
4. It is an authoritative text suitable for use in the teaching and study of animal law at a growing number of universities in Britain and other countries (Sankoff 2008).
5. Its publication was timely in that an easy to read account of animal law in Britain was available to support public consultation prior to and after the release of the *Draft Animal Welfare Bill* launched by the Department for Environment, Food and Rural Affairs in 2004 (Department for Environment, Food and Rural Affairs 2004).
6. It makes suggestions for reform to improve animal protection law in Britain, including advocating the adoption of a duty of care in the Protection of Animals Acts as was later introduced in the *Animal Welfare Act 2006* (Department for Environment, Food and Rural Affairs 2006).
7. It demonstrates that, even though the progress of reform is slow, incremental changes in the law can achieve enhanced protection for animals.
8. It shows that current animal protection law in Britain had its roots in the nineteenth century.
9. It describes how the previously existing sovereignty of the national Parliament at Westminster has been eroded by the accession of the UK to the European Community, the operation of the World Trade Organisation, and the devolution in the UK of legislative power to Scotland and Wales.
10. It illustrates the hurdles to animal welfare arising from the existence of trade liberalisation policies within the European Community and the World Trade Organisation.
11. It compares the legal concept of cruelty and the scientific discipline of animal welfare.
12. It highlights the importance of scientific evidence in informing animal welfare legislation and argues for the application of the precautionary principle to give animals the benefit of doubt where there is uncertainty.
13. It demonstrates the importance of pressure groups in achieving legislative reforms for animals.
14. It makes a detailed analysis of legal provisions.

15. It shows that provisions in legislation can provide a guide as to how animals should not be treated as illustrated by offences in the cruelty legislation.
16. It examines the importance of licensing, certification, and regulation in achieving improved animal welfare and providing a buffer to the unfettered treatment of animals as property under common law.
17. It argues that proper enforcement needed to make animal protection legislation effective involves much more than just prosecuting offenders.
18. It highlights the important role played by the RSPCA and SSPCA as charitable organisations in providing protection to animals.
19. It constitutes a valuable reference on animal law in Britain for law reformers, including bureaucrats and politicians, and for those responsible for the care and use of animals in a variety of situations.
20. It provides guidance and ideas for people in other jurisdictions involved in the introduction of or reform to laws to improve animal protection.

Thus, there has been enormous scope for *Animal Welfare Law in Britain* to be influential in many different ways during the decade since its publication. Indeed, as observed by Margaret Cooper (2002, p. 258) in her review of the book:

> In conclusion, this is a classic text that will be a standard reference for all who work, practise, study and teach in this field. It pulls together the many aspects that influence animal welfare legislation and sets out the origins and development of these in a well-referenced and scholarly work that is, nonetheless, very readable. Mike Radford has done us a great service in weaving these matters together with meticulousness based on extensive reading of the sources.

9.4 Potential for the Book

The potential for *Animal Welfare Law in Britain* (Radford 2001) is closely related to many of the factors listed above in relation to its influence. However, the potential for the book would be greatly enhanced by the publication of a revised edition that takes into account changes to the law in Britain resulting from enactment of the *Animal Welfare Act 2006* (Department for Environment, Food and Rural Affairs 2006). The 2006 Act was designed to modernise and consolidate 21 separate Acts including the *Protection of Animals Act 1911* into a single statute covering the welfare of domestic and captive animals in England and Wales (Department for Environment, Food and Rural Affairs 2004). Enactment of the *Animal Welfare Act 2006* extended to non-farmed animals a positive duty to ensure their welfare as recently adopted in relation to farmed animals under the *Agriculture (Miscellaneous Provisions) Act 1968*.

Even though *Animal Welfare Law in Britain* (Radford 2001) essentially covers the development and contents of animal protection law in Britain prior to 2001, it does provide a reliable source of information and ideas of potential value to a range of people in both Britain and other jurisdictions wishing to understand more comprehensively about the role and scope of law in improving the lives of animals.

It shows how animal protection law has evolved in Britain, the first national jurisdiction in the world to adopt such legislation, and indicates how it might be reformed further in the future. Indeed, the book advocated the introduction of a positive duty of care as already applying to farm animals into the general animal protection legislation as was done subsequently in the *Animal Welfare Act 2006*. If and until a revised edition of *Animal Welfare Law in Britain* (Radford 2001) is published, the book needs to be viewed in the light of more recent developments in animal protection law in Britain. Nevertheless, the book provides a very useful source for increasing understanding and generating ideas that would be of value for animal activists and law reformers in other jurisdictions wishing to introduce or reform legislation to improve animal protection and ensure it is properly enforced. There is little value in "reinventing the wheel", and much can be gained by looking critically at the experience of Britain in developing legislation to prevent cruelty to animals and improve their welfare. Indeed, the British *Draft Animal Welfare Bill* launched in 2004 incorporated reforms already undertaken in some other countries, notably New Zealand and Sweden (Department for Environment, Food and Rural Affairs 2004, p. 46), and in a similar way, *Animal Welfare Law in Britain* (Radford 2001) read in conjunction with the *Animal Welfare Act 2006* can provide valuable background information to be considered in developing or reforming animal protection legislation in other countries, particularly as the book does much more than just state black letter law. More broadly, the book continues to have potential for use by a range of people including teachers and students of animal law, animal activists, law reformers, and the owners or users of animals of various kinds, both in Britain and overseas, wishing to create better legal protection for animals.

9.5 Some Other Books

Even though now largely of historical value, mention should be made of *Animals and the Law* by T.G. Field-Fisher published in 1950 by the Universities Federation for Animal Welfare (UFAW) and revised in 1964 (Field-Fisher 1950, 1964). The book was described by Cooper (2002, p. 257) as a splendid book that, even though compact, provided a succinct, comprehensible, and accurate summary of animal law in Britain. Moreover, the book was a valuable guide at the time not only for lawyers but for anyone wishing to know more about animal law.

The book *Animal Rights: Current Debates and New Directions* edited by Professors Cass Sunstein and Martha Nussbaum of the University of Chicago was published in 2004 (Sunstein and Nussbaum 2004). It comprises an introduction on animal rights by Cass Sunstein, 14 essays by well-known contributors, and a biographic essay. The book is divided into two parts, namely Part I: Current Debates and Part II: New Directions, and it explores legal and political issues underlying the campaign for animal rights and the opposition to it. The various authors put forward a range of perspectives on animal rights and animal welfare, and these provide food for thought that hopefully will be manifested eventually in improved legal protection for animals. The book is aimed at bringing a new clarity to the animal rights debate with focus on the practical and at charting some new directions for both theory and practice. Among the issues considered are the drawing of a line between species, application of the capabilities approach to animal rights, and the matter of legal standing for animals.

In 2005, the World Organisation for Animal Health (OIE) published a comprehensive book entitled *Animal Welfare: Global Issues, Trends and Challenges* (World Organisation for Animal Health 2005). The book was coordinated by Drs David Bayvel, Sira Abdul Rahman, and Andrea Gavinelli, and it comprises a large number of individual contributions by leaders in a range of subject areas related to animal welfare. Previously, in 2002, the World Organisation for Animal Health (OIE) had received "a unanimous mandate from its 167 Member Countries to become the leading international organisation in the field of animal welfare" (Vallat 2005). The publication of *Animal Welfare: Global Issues, Trends and Challenges* was intended to be a contribution by OIE towards its commitment to leadership in science-based animal welfare at the international level, and to communicate with Member Countries, the private sector, and society in general. The book is divided into five sections, namely:

Science-Based Evaluation of Animal Welfare
Global Perspectives: OIE Regions and Member Countries
Global Perspectives: Stakeholders
OIE Animal Welfare Standards
Strategic Animal Welfare Issues

The conclusion of the book consists of a summary paper by Dr David Bayvel, Director of Animal Welfare at the New Zealand Ministry of Agriculture and Forestry, in which he explores the historical context, international considerations, and future direction for the use of animals in agriculture and science (Bayvel 2005). In a review of the book, Dorothy McKeegan (2007, pp. 529–530) regarded the publication by OIE as more like a series of related journal articles than a book as such and noted that the contributions varied considerably in style and technicality. Nevertheless, she concluded that the book contained an enormous amount of information that made it a useful resource for anyone interested in animal welfare and the way it was developing on a global scale. Moreover, the book contains valuable background information for use by OIE in the creation of international animal welfare standards (Petrini and Wilson 2005).

The book *Animal Suffering and the Law* (Eadie 2009) is intended to provide a broad understanding of the role of law in animal protection at various jurisdictional levels. The book covers a wide range of issues at national, regional, and international level that are relevant to the legal protection of animals against suffering inflicted by humans. Among the national issues discussed are the limitations in existing legislation designed to protect animals, problems with codes of practice, and the role of animal protection organisations. The regional matters covered include the substantive animal protection legislation in the European Union and suggestions for reform, the contributions of Eurogroup towards improving the legal protection of animals in the European Union, and conflict between the World Trade Organisation (WTO) and European Community law on animal welfare. At an international level, the issues considered include the existence and effectiveness of international conservation and environmental agreements that provide indirect protection to animals, the provisions of International Humanitarian Law designed to protect the civilian population that also give indirect protection to animals, and a consideration of lessons from the evolution of International Human Rights Law for the future development of international legal protection for animals.

In 2008 *Handbook of Australian Animal Cruelty Law* by Dr Malcolm Caulfield, Legal Counsel for Animals Australia, was published (Caulfield 2008). The book covers Australian state, territory, and commonwealth legislation and examines its application to the use of animals in agriculture, live animal export, animal slaughter, wild and feral animal culling, and the use of animals in scientific research and teaching. It deals also with the matter of enforcement and includes some case studies involving live export, mulesing, and intensive piggeries. The book is critical of the inadequacy of the law in Australia to properly protect animals from large-scale suffering in agribusiness operations as well as in other situations. Grants from Voiceless and the Victoria Law Foundation enabled a significant number of copies of the book to be distributed free to students and others, and the book should facilitate greater awareness of animal protection issues in Australia.

Another book *Animal Law in Australasia: A New Dialogue* edited by Peter Sankoff of the University of Auckland in New Zealand and Steven White of Griffith University in Australia was published in 2009 (Sankoff and White 2009). The book contains contributions from 15 separate authors including the two editors, and it is divided into four parts, namely:

Part I: Core Concepts in Animal Law
Part II: Jurisprudential Challenges in Animal Law
Part III: Animal Welfare in Australasia: Specific Challenges for the Region
Part IV: Looking Abroad and into the Future

The book is intended (Sankoff and White 2009, p. 4) to stimulate dialogue regarding the legal relationship between humans and animals and break a silence that so easily maintains the status quo. The book should make a valuable contribution to the growing animal law movement in Australasia as well as inspire a greater involvement by legal scholars in animal protection issues.

9.6 Overview

In *Animal Welfare Law in Britain: Regulation and Responsibility*, Mike Radford (2001) traces the evolution of animal protection law in Britain from its beginning in 1822–2001, and he examines the changing emphasis from legislation to prevent animal cruelty to legislation in more recent years to improve the welfare of animals so that their needs are met. Indeed, there has been a shift from prevention of cruelty to animal welfare legislation in a number of jurisdictions including New Zealand, Australia, and Britain. Nevertheless, a great deal needs to be done to ensure that the welfare of various categories of animals including intensively farmed animals is appropriately provided for. In this the science of animal welfare has an important role to play in properly informing reform. However, there is substantial political opposition to reform that is driven by vested economic interests. In order to help overcome such obstacles to reform, there needs to be greater transparency and media exposure, coupled with scientific evidence, that enables the public to become more aware about the way animals are being treated, including in intensive farming situations. Voters can then put informed pressure on government to provide better legal protection for the animals involved. At present, for instance, a huge number of intensively farmed animals such as pigs and poultry are subjected to practices and conditions that would amount to cruelty if the animals were companion animals such as a dog or a pet bird. Indeed, it appears (Eadie 2009, pp. 47–56) that generally only companion animals are afforded broad protection by the law from human-inflicted suffering. However, there exists a capacity for the law to provide even greater protection for companion animals as well as others by means of licensing, certification, and registration as suggested by Radford (2001, pp. 291–307).

The need for legal protection of animals is universal. A useful contribution towards this is being made by the World Organisation for Animal Health (OIE) in developing what are minimum international standards of animal welfare (Petrini and Wilson 2005, pp. 665–671). Ultimately, however, each nation has to enact domestic animal protection legislation and ensure it is effectively enforced. In this, the experience of nations such as Britain whose jurisdictions already have in place well developed and tested legislation to protect animals, despite its inadequacies, can be used as a guide in the introduction or reform of legislation in other jurisdictions. Moreover, the book *Animal Welfare Law in Britain* (Radford 2001) can provide valuable background information and assist in the generation of ideas for the enactment of animal protection legislation in any jurisdiction. In some developing countries, the issue of poverty is a major impediment to improving animal protection, and when this is overcome, many people, who previously found it difficult to survive themselves, will be in a better situation to be able to provide proper welfare for their animals. However, it is necessary through education for people all over the world to be made aware that animals are sentient beings capable of suffering and that it is immoral for humans to inflict suffering upon them.

to animal welfare legislation in a number of jurisdictions including New Zealand, Australia, and Britain. Nevertheless, a great deal needs to be done to ensure that the welfare of various categories of animals including [illegible] farmed animals is appropriately provided for [illegible] so that the science of animal welfare has an important role to play in [illegible]. However, there [illegible]

[illegible]

Chapter 10
Conclusions

Abstract The chapters covered in this book identify that there is a powerful animal welfare and rights movement that is challenging the existing animal management systems, which bring major profits to industry and which are inadequately protected by legislative frameworks. Improvements are frequently advocated through the use of encouragement to animal managers, the 3Rs and the Five Freedoms.

This book examines the contents, influence, and potential of a personal selection of modern books published over the previous 50 years or so that have relevance to improving the welfare of animals affected by the control or actions of humans. The books selected cover both earlier seminal works of the period and specific subject disciplines that include history, science, ethics, politics, and law.

The first three selected books are earlier seminal works of the modern period. In 1959, William Russell and Rex Burch published *The Principles of Humane Experimental Technique* in which they advocated the use of replacement, reduction, and refinement for removing or diminishing the inhumanity of experimental techniques involving animals. These are commonly referred to as the Three Rs, and they are now well known, even though they are not applied as effectively as they could be. Five years later in 1964, Ruth Harrison published *Animal Machines: The New Factory Farming Industry* that exposed the suffering of animals in modern intensive farming operations and resulted in the establishment in 1965 of the Brambell Committee in Britain to investigate such operations, and this led eventually to what are now commonly known as the Five Freedoms for animals. In 1975 Peter Singer published *Animal Liberation: Towards an End to Man's Inhumanity to Animals* in which he showed the immorality of the infliction of suffering on animals by humans and the existence of discrimination against animals by humans known as speciesism. He illustrated such immorality and speciesism by describing examples of human conduct towards animals in laboratory situations and intensive farming, and he advocated the adoption by humans of a vegetarian diet as a means of reducing animal suffering. The book by Singer has been influential in changing

E.N. Eadie, *Understanding Animal Welfare*, Animal Welfare 13,
DOI 10.1007/978-3-642-30577-1_10, © Springer-Verlag Berlin Heidelberg 2012

human attitudes towards animals, but much remains to be done to put his philosophy more widely into practice.

Subsequently, five selected books that deal with historical, scientific, activist, political, and legal aspects of animal welfare are considered. In 1989 Richard Ryder published *Animal Revolution: Changing Attitudes Towards Speciesism* that describes historical developments in the animal movement and puts improvements in animal protection into a historical context. A practical account of the developing science of animal welfare was presented in 1995 by John Webster in his book *Animal Welfare: A Cool Eye Towards Eden*. In 1998, Peter Singer published *Ethics into Action: Henry Spira and the Animal Rights Movement* that is a biography of the animal protection campaigner Henry Spira, who achieved some major reforms by means of direct action for both experimental and farmed animals in the USA between the mid-1970s and the mid-1990s. The politics of animal protection was analysed by Robert Garner in his book *Political Animals: Animal Protection Politics in Britain and the United States* published in 1998 in which he examines the difficulties associated with achieving improvements in animal welfare by political means, in particular, for experimental and farmed animals. In 2001 Mike Radford published *Animal Welfare Law in Britain: Regulation and Responsibility* that traces the evolution of animal protection law in Britain and makes suggestions for further improvements in the law. The book has the potential to provide ideas for the introduction or development of animal protection law in other jurisdictions.

Each of the selected books has relevance to some aspect of improving animal welfare, and the messages contained in them can be used as part of an integrated approach to enhancing animal welfare around the globe. Indeed, animal protection involves many issues and disciplines, and these need to be integrated and human interference in animals' ecosystems and lives minimised to reduce their suffering at the hands of humans. The Three Rs of Russell and Burch for reducing the inhumanity of animal experimentation and testing need not only to be known but to be applied conscientiously and effectively. This involves sound education in animal ethics, the will and capacity of experimenters to search for alternatives, the proper use of statistics in experimental design if animals are to be used, the adoption of compassionate attitudes towards animal suffering, and the effective enforcement of legal provisions. In the case of farmed animals, the Five Freedoms that developed from the Brambell Report, a consequence of the book by Ruth Harrison, need to be understood and applied. Indeed, farm operators, animal activists, politicians, and other law reformers should have some appreciation of animal welfare science as expounded by John Webster so that the treatment of animals and improvements in their welfare are based on credible evidence. In cases where there is scientific uncertainty, the precautionary principle should be applied so animals are given the benefit of the doubt.

The selected earlier seminal works of William Russell and Rex Burch, Ruth Harrison, and Peter Singer have had substantial influence in improving animal welfare, and they are likely to continue to do so well into the future. The Three Rs proposed by Russell and Burch in 1959 are now well known, whether arising directly or indirectly from the book, and they are incorporated into animal welfare education, practised by scientists using animals in experimentation, and are enshrined in national and regional policy and legislation. They are also a guiding principle for the World Organisation for Animal Health (OIE) in developing animal welfare standards to provide a minimum of protection for animals on a global basis. With education, commitment, and legal enforcement, the Three Rs are likely to play an important role towards finding alternatives to replace animals in experimentation and contributing to more humane treatment for animals that are used in experimentation. The book by Ruth Harrison published in 1964 provided the first public exposure of animal welfare issues associated with intensive farming. It was the catalyst for the setting up in 1965 of the Brambell Committee in Britain from which the Five Freedoms for animals were eventually developed, and these are now well known. The Five Freedoms form part of animal welfare education as, for instance, in the *Concepts in Animal Welfare* syllabus developed by the World Society for the Protection of Animals (WSPA) in association with Bristol University for use around the globe. Also, the Five Freedoms influence public policy, provide a yardstick for assessing the provisions in animal protection legislation, and are a guiding principle of the OIE in developing international standards for animal welfare. These standards are likely to be raised incrementally over time assisted by developments in the science of animal welfare. Even though some of the improvements relating to the welfare of intensively farmed animals proposed by Harrison have been slow to be put into practice, she started a movement that with perseverance has eventually brought about reforms. Indeed, without the transparency regarding the treatment of intensively farmed animals initiated by Harrison coupled with ongoing activism and perseverance, it is unlikely that the reforms that have been achieved would have come about. The publication in 1975 of *Animal Liberation* by Peter Singer reinforced the findings of Ruth Harrison. The accessibility of *Animal Liberation* to the community at large has played an enormously important role in providing a moral basis for opposition to the way humans treat animals, exposed widely what was actually happening to animals in experimentation and intensive farming, changed public sentiment towards animals, and resulted in the formation of animal protection organisations in many countries. This provided the constituency necessary to influence public policy and resulted in legal enactments to abolish various morally unacceptable practices that caused suffering to animals. These included the banning in various jurisdictions of veal crates, sow stalls, and battery cages for layer hens.

The five selected books dealing with specific subject disciplines are influential in that they play an important role in relation to an integrated approach to improving animal protection. No single discipline or means on its own is likely to achieve the changes necessary to provide adequate improvements in animal welfare. The book by Richard Ryder published in 1989 on the history of the animal movement put the challenge of improving animal protection into a historical perspective and showed the difficulties involved but gives some inspiration that with perseverance and credible campaigning improvements in the welfare of animals can be achieved. The book by John Webster published in 1995 on the science of animal welfare demonstrates a more rigorous basis for assessing animal welfare, particularly by consideration of the Five Freedoms for animals he helped develop, and this provides the credible evidence that is necessary to properly inform public policy and result in improved legislative protection for animals. However, it is not sufficient to just have the evidence. It is necessary as described by Robert Garner in his book on the politics of animal protection published in 1998 to understand how public policy decisions are made and, for a sufficiently large constituency to develop, to exert pressure on and get the attention of politicians, who have many different issues to consider at any time. This is seen in relation to the recent media exposure of the suffering experienced by exported Australian cattle at Indonesian abattoirs. It is interesting to note also that the organisation Animals Australia that drew the attention of the media to Indonesian abattoir abuse was originally formed under the direct influence of the book *Animal Liberation* together with the personal involvement of Peter Singer. Moreover, the public exposure of morally indefensible practices regarding the way animals are treated plays an important role in bringing about change directly as described by Singer in his biography of activist Henry Spira published in 1998. In other cases changes in public policy can result in improvements in animal welfare legislation. To assist in making such legislation appropriate and effective, valuable ideas regarding legislative enactment and enforcement can be gained by reference to the book on animal welfare law by Mike Radford published in 2001. Moreover, bureaucrats rather than politicians are usually responsible for the details of any legislative change, and they are more likely than politicians to have read books on animal welfare such as that of John Webster. Each of the books selected contributes directly or indirectly to the diverse and interrelated means that play a role in the achievement of proper protection for animals against human-inflicted suffering. The earlier selected seminal works have already had an important influence in the improvement of animal protection as discussed in this book, whereas the more recent selected books have considerable potential to influence the improvement of animal protection in the future. No one selected book alone is going to provide all the answers, but taken together, they contain the ingredients for an integrated approach to improving animal welfare.

The creation of transparency as in the books of Ruth Harrison and Peter Singer is needed to inform the media and the public, and this in turn provides the necessary pressure to influence politicians and change public policy. However, it is desirable for public policy to be informed by credible evidence as provided by animal welfare science. Moreover, public policy needs to be incorporated into properly enacted and enforced legal provisions to be effective. Thus, the lessons from the various books selected need to be used in combination to achieve proper animal protection. Moreover, the selected books have had or can have influence on the various means for achieving improved animal protection. For instance, the Three Rs of Russell and Burch are used in animal welfare education, are incorporated into public policy and legislation, and result in the development of alternatives to animal use as well as reduced animal suffering where animals continue to be used in experimentation. The book of Ruth Harrison was the catalyst for media exposure of intensive farming practices, influenced public opinion, gained the attention of politicians, resulted in the development of the Five Freedoms for animals, and ultimately resulted in legislative changes nationally and regionally to abolish certain practices such as veal crates, sow stalls, and battery cages. The book *Animal Liberation* by Peter Singer, who previously had been influenced by Ruth Harrison's book, has also influenced public opinion and was a major catalyst for the development of the modern animal movement including the formation of numerous animal protection organisations in various countries that play an important role in changing public sentiment towards animals. This in turn can result both in direct reforms and in changes to public policy followed by legislative enactments to improve animal protection. The creation of transparency leads also to changes in consumption habits such as the demand for free-range eggs and the adoption of a vegetarian or vegan diet.

It is good that moves are underway to find a satisfactory mechanism by which books in the great libraries of the world such as that of Harvard are made available online. The achievement of this would be important as many people now rely largely on information available online, and it would make all the books selected or discussed in this book more readily available to a larger readership. Even though some people access mainly published papers, it is through books that various published papers are brought together to provide a more integrated picture. In a similar way, the present work covers a number of books that taken together provide the ingredients of an integrated approach to improving animal welfare and rights. Moreover, the books selected or considered provide a broad appreciation of a diverse range of animal protection issues and disciplines. This is important as these various components need to be seen in perspective and used to maximum advantage in improving the lives of animals.

It is essential to have transparency regarding the way animals are treated so that members of the public can make informed decisions in their purchasing choices as well as exert pressure on government for reform. Activists need to have credibility with politicians, and the animal movement has to develop a sufficiently large constituency to influence government to improve animal welfare as concluded by Robert Garner. Transparency also facilitates reform by direct action based on ethical persuasion as achieved by Henry Spira as described by Peter Singer. Such reforms can be cemented subsequently into legal regulation to avoid the recurrence of ethically unacceptable practices. Moreover, transparency accompanied by political pressure on government can be the catalyst for reform of the law. The need for proper animal protection is universal, and law reform to improve the lives of animals adopted in one jurisdiction can be used as a guide for reform in other jurisdictions. Indeed, it is desirable for a watching brief to be kept of positive legal developments for animals everywhere. However, any legal provisions enacted have to be properly enforced to make them effective as observed by Mike Radford.

Greater recognition of the immorality and continuing existence of the infliction of suffering on animals by humans can lead more people to change their behaviour as when they adopt a vegetarian or vegan diet as advocated by Peter Singer. However, the eating of meat and other animal products can be expected to continue, and indeed it is increasing on a global basis, so there is need for transparency coupled with meaningful and reliable food product labelling so that consumers are able to make informed choices that reduce or eliminate animal suffering. Moreover, for more people to make ethical choices, there has to be greater awareness of animal welfare issues as exposed by Ruth Harrison, Peter Singer, and John Webster. Besides, it is not only experimental and farmed animals that experience human-inflicted suffering but many other animals also in a range of different situations, so there needs to be a change in attitude generally by humans towards animal suffering. In this education and transparency have important roles to play.

History shows that reform to improve the lives of animals can be slow as revealed by Richard Ryder, and seen in the present work, but that with perseverance and dedication reform can be achieved. Many different disciplines including philosophy, science, activism, politics, and law each have their part to play in an integrated approach to creating a better world for animals, and this has to be seen as a global challenge because human-inflicted animal suffering is a universal problem. The books selected for detailed consideration in this work, as well as the related books mentioned or discussed, have been influential and offer potential in the future regarding many important issues that are relevant in the journey towards the creation of better welfare for the animals whose lives are controlled or affected by the actions of humans.

References

Animal Care and Protection Act (2001) Reprint no 2A, reprinted as in force on 6 November 2003, Queensland

Animals Today (1995) Animal liberation – the book that has changed perceptions about animals. Animals Today, February–April, pp 8–12

Appleby M (1999) What should we do about animal welfare? Blackwell, Oxford

Arey D (2004) Practical alternatives to battery cages for laying hens: case studies from across the European Union. Compassion in World Farming Trust, Petersfield

Arey D, Brooke P (2006) Animal welfare aspects of good agricultural practice: pig production. Compassion in World Farming Trust, Petersfield

Armstrong S, Botzler R (eds) (2003) The animal ethics reader. Routledge, London

Australian Animal Welfare Strategy Education and Training Working Group (2008) Australian Animal Welfare Strategy Education and Training Stocktake – Executive Summary and Background to Report, Australian Animal Welfare Strategy Education and Training Working Group, February 2008

Balcombe J (2010) Second nature – the inner lives of animals. Macmillan, New York

Ball M, Friedrich B (2009) The animal activist's handbook – maximizing our positive impact in today's world. Lantern, New York

Balls M (2009) The three Rs and the humanity criterion: an abridged version of the principles of humane experimental technique by WMS Russell and RL Burch (1959). The Fund for the Replacement of Animals in Medical Experiments, Nottingham

Balluch M (2006) How Austria achieved a historic breakthrough for animals. In: Singer P (ed) In defense of animals – the second wave. Blackwell, Oxford, pp 157–166

Bayvel A (2005) The use of animals in agriculture and science: historical context, international considerations and future direction. In: Animal welfare: global issues, trends and challenges, World Organisation for Animal Health, Paris, pp 791–797

Boyd K (2010) Book review of 'The three Rs and the humanity criterion: an abridged version of principles of humane experimental technique by WMS Russell and RL Burch (1959)' by M. Balls (2009), Animal welfare. Universities Federation for Animal Welfare, Wheathampstead, p 197

Brambell R (1965) Report of the Technical Committee to enquire into the welfare of animals kept under intensive livestock husbandry systems (Brambell Report), Command Paper 2836, Her Majesty's Stationery Office, London

Broom D (2001) Animal welfare at Cambridge, interview of Professor Don Broom by Glenys Oogjes. Anim Today 9(2):24–26

Broom D, Fraser A (2007) Domestic animal behaviour and welfare. CAB, Wallingford

Brown L (ed) (1993) The new Shorter Oxford English Dictionary, vol 2: N-Z. Clarendon, Oxford

E.N. Eadie, *Understanding Animal Welfare*, Animal Welfare 13,
DOI 10.1007/978-3-642-30577-1,

Budd W (1988) Hear the other side – a history of the Royal Society for the Prevention of Cruelty to Animals in South Australia 1875–1988. Investigator Press, Hawthorndene

Carson R (1964) Foreword. In: Harrison R (ed) Animal machines – the new factory farming industry. Vincent Stuart, London, pp vii–viii

Caulfield M (2008) Handbook of Australian Animal Cruelty Law. Animals Australia, North Melbourne

Coetzee J (2010) Foreword. In: Balcombe J (ed) Second nature – the inner lives of animals. Macmillan, New York, pp ix–xii

Commonwealth of Australia (2003) Australian code of practice for the care and use of animals for scientific purposes. Commonwealth of Australia, March 2003

Cooper M (2002) Book review of 'Animal welfare law in Britain' by M. Radford (2001), Animal welfare. Universities Federation for Animal Welfare, Wheathampstead, pp 255–258

Cox J (2007) Industrial animal agriculture – part of the poverty problem. World Society for the Protection of Animals, London

Dawkins M, Bonney R (eds) (2008) The future of animal farming: renewing the ancient contract. Blackwell, Oxford

de Boo J, Knight A (2005) "Concepts in animal welfare": a syllabus in animal welfare science and ethics for veterinary schools. J Vet Med Educ 32(4):451–453

de Fraga C (2000) Vale Ruth Harrison 5.6.1920-13.6.2000. Anim Today 8(3):26–27

de Fraga C (2002) Germany: laying hens to spread their wings. Anim Today 10(1):5

Department for Environment, Food and Rural Affairs (2004) Launch of the Draft Animal Welfare Bill, Animal Welfare Division, Department for Environment, Food and Rural Affairs, Presented to Parliament July 2004, UK

Department for Environment, Food and Rural Affairs (2006) Animal Welfare Act, Department for Environment, Food and Rural Affairs, UK. http://www.defra.gov.uk, 9 Nov 2006

Department of Agriculture, Fisheries and Forestry (2005) The Australian Animal Welfare Strategy. Department of Agriculture, Fisheries and Forestry, Australian Government, Canberra

Duncan I, Hawkins P (eds) (2010) The welfare of domestic fowl and other captive birds. Springer, Heidelberg

Dunlop F (2008) Shark's Fin and Sichuan Pepper: a sweet-sour memoir of eating in China. Ebury, London

Dwyer C (ed) (2008) The welfare of sheep. Springer, Heidelberg

Eadie E (2009) Animal suffering and the law – national, regional, and international. Seaview, West Lakes

Eadie E (2011) Education for animal welfare. Springer, Heidelberg

Egerton J (1992) Ministerial review of the Prevention of Cruelty to Animals Act 1979 and Regulations. Public discussion paper, vol 1, NSW Agriculture, November 1992

Eurogroup for Animals (2008) Responsible retailing – putting animal welfare at the heart of your food products supply chain. Eurogroup for Animals, Brussels

Eurogroup for Animal Welfare & RSPCA International (2001) The European Union accession and animal welfare – an introduction to European animal welfare legislation. Eurogroup for Animal Welfare/RSPCA International, Brussels/Horsham

Eurogroup for Animal Welfare (2003) Annual report 2002 – Eurogroup for Animal Welfare: animal protection through legislation, Brussels, April 2003

Eurogroup for Animal Welfare (2004) Analysis of major areas of concern for animal welfare in Europe, Eurogroup for Animal Welfare, Brussels, July 2004

Ferguson S, Doyle M (2011) A bloody business. Four Corners, ABC Television, 30 May 2011

Festing M (2009) Fifty Years after Russell and Burch, toxicologists continue to ignore genetic variation in their test animals. Alt Lab Anim 37(1):1–5

Festing S (2010) Don't waste the animals. New Scientist, 5 June 2010, pp 22–23

Field-Fisher T (1950) Animals and the law. Universities Federation for Animal Welfare, Wheathampstead

Field-Fisher T (1964) Animals and the law. Universities Federation for Animal Welfare, Wheathampstead, Revised edition
Finsen L, Finsen S (1994) The animal rights movement in America. Twayne, New York
Fisher A (2010) Book review of 'Improving animal welfare: a practical approach' edited by T. Grandin (2009), Animal welfare. Universities Federation for Animal Welfare, Wheathampstead, p 546
Foer J (2009) Eating animals. Hamish Hamilton – Penguin, London
Francione G (1995) Animals, property, and the law. Temple University Press, Philadelphia, PA
Francione G, Garner R (2010) The animal rights debate – abolition or regulation? Colombia University Press, New York
Fraser A (1980) Farm animal behaviour. Bailliere Tindall, London
Fraser A, Broom D (1990) Farm animal behaviour and welfare. Bailliere Tindall, London
Fraser D (2005) Book review of 'Animal welfare: limping towards Eden' by J. Webster (2005), Animal welfare. Universities Federation for Animal Welfare, Wheathampstead, pp 391–392
Garner R (1998) Political animals – animal protection politics in Britain and the United States. Macmillan, London
Gauthier C, Griffin G (2001) Book review of 'Animal revolution: changing attitudes towards speciesism' by R.D. Ryder (2000), Animal welfare. Universities Federation for Animal Welfare, Wheathampstead, pp 108–110
Grandin T (ed) (2010) Improving animal welfare – a practical approach. CAB, Wallingford
Harrison R (1964) Animal machines – the new factory farming industry. Vincent Stuart, London
Harrison R (1996) Book review of 'Animal welfare: a cool eye towards Eden' by John Webster (1995), Animal welfare. Universities Federation for Animal Welfare, Wheathampstead, pp 452–455
Harrison R (2000) Book review of 'What should we do about animal welfare?' by M. Appleby (1999)', Animal welfare. Universities Federation for Animal Welfare, Wheathampstead, pp 92–94
Hart L (2004) Book review of 'From guinea pig to computer mouse: alternative methods for a progressive, human education, 2nd edition' by N. Jukes and M. Chiuia (2003), Animal welfare. Universities Federation for Animal Welfare, Wheathampstead, pp 255–257
Hart L, Wood M, Weng H-Y (2005) Effective searching of the scientific literature for alternatives: search grids for appropriate databases. Three Rs special issue, Animal welfare, vol 14, issue 4. Universities Federation for Animal Welfare, Wheathampstead, pp 287–289
Hughes J (1995) Can the 3R's – reduction, refinement, replacement – free laboratory animals? Interview by Judy Hughes with Professor Michael Balls. Anim Today 3(3):14–15, 32
Innes S (2009) Free-range eggs crack luxury hotel market. The Advertiser, Adelaide, April 10, p 15
Jasper J, Nelkin D (1992) The animal rights crusade – the growth of a moral protest. Free Press, New York
Jennings S (1964) Preface. In: Harrison R (ed) Animal machines – the new factory farming industry. Vincent Stuart, London
Jones T (2010) Jonathan Safron Foer joins Lateline. Lateline Program, ABC Television, 26 Mar 2010
Jukes N, Chiuia M (2003) From guinea pig to computer mouse: alternative methods for a progressive, humane education, 2nd edn. International Network for Humane Education (InterNICHE), Leicester
Kaliste E (ed) (2004) The welfare of laboratory animals. Springer, Dordrecht
Kirkwood J (2005) Editorial. Three Rs special issue, Animal welfare, vol 14, issue 4. Universities Federation for Animal Welfare, Wheathampstead
Marchant-Forde J (ed) (2009) The welfare of pigs. Springer, Heidelberg
Mathews D (2007) Committed – a Rabble-Rouser's Memoir. ATRIA, New York

McKeegan D (2007) Book review of 'Animal welfare: global issues, trends and challenges' published by the World Organisation for Animal Health (2005), Animal welfare. Universities Federation for Animal Welfare, Wheathampstead, pp 529–530
Mepham B (2011) Book review of 'The animal rights debate – abolition or regulation?' by G. Francione and R. Garner (2010), Animals. 1. 200–204
Morton D (2004) Experimental procedures: general principles and recommendations. In: Kaliste E (ed) The welfare of laboratory animals. Springer, Dordrecht, pp 81–115
National Animal Ethics Advisory Committee (2004) National Animal Ethics Advisory Committee 2003 Annual Report, National Animal Ethics Advisory Committee, Wellington
Newby J (2009) Elephants. Catalyst, ABC Television, 2 Apr 2009
Newkirk I (2005) Making kind choices. St Martin's Griffin, New York
Noble D (2004) Virtual organs. Interview of Professor Denis Noble by Robyn Williams, Broadcast on 'The Science Show', ABC Radio National, 6 Nov 2004
Oogjes G (1998) Review of 'Ethics into action – Henry Spira and the animal rights movement' by Peter Singer. Anim Today, August–October, p 28
O'Sullivan S (2006) Transparency and animal research regulation: an Australian case study. Animal Liberation Philos Policy J 4(1)
Pacelle W (2011) The bond – our kinship with animals, our call to defend them. Harper Collins, New York
Petrini A, Wilson D (2005) Philosophy, policy and procedures of the World Organisation for Animal Health for the development of standards in animal welfare. In: Animal welfare: global issues, trends and challenges, World Organisation for Animal Health, Paris, pp 665–671
Phillips C (2008) 'From the director', CAWE update, The Centre for Animal Welfare and Ethics, The University of Queensland, November 2008
Phillips C (2009) The welfare of animals – the silent majority. Springer, Heidelberg
Phillips C (2011a) Animal welfare series preface. In: Eadie EN (ed) Education for animal welfare. Springer, Heidelberg, pp v–vii
Phillips C (2011b) Personal communication sent to the author by Professor Clive Phillips, 23 June 2011
Phillips C, McCulloch S (2005) Student attitudes on animal sentience and use of animals in society. J Biol Educ 40:17–24, Winter 2005
Radford M (1999) Book review of 'Political animals: animal protection politics in Britain and the United States' by R. Garner (1998), Animal welfare. Universities Federation for Animal Welfare, Wheathampstead, pp 303–304
Radford M (2001) Animal welfare law in Britain – regulation and responsibility. Oxford University Press, Oxford
Regan T (1983) The case for animal rights. University of California Press, Berkeley, CA
Regan T (1985) The case for animal rights. In: Singer P (ed) In defence of animals. Blackwell, Oxford, pp 13–25
Regulations under the Prevention of Cruelty to Animals Act (1985) No.10 of 2000, South Australia
Rochlitz I (ed) (2005) The welfare of cats. Springer, Dordrecht
Rollin B (1981) Animal rights and human morality. Prometheus, New York
Rollin B (2003) Foreword. In: Armstrong S, Botzler R (eds) The animal ethics reader. Routledge, London, pp xi–xiii
Rowan A (1984) Of mice, models and men. State University of New York Press, Albany, NY
RSPCA Australia (2008) Policies. RSPCA Australia, Deakin West, ACT, 2008 edition
RSPCA South Australia (2009) Choose wisely. RSPCA South Australia Annual Report 2009, RSPCA South Australia, p 6
RSPCA United Kingdom & Eurogroup for Animal Welfare (2001) Conflict or concord? Animal Welfare and the World Trade Organisation, RSPCA & Eurogroup for Animal Welfare
Rushen J, de Passillé A, von Keyserlingk M, Weary D (2008) The welfare of cattle. Springer, Dordrecht
Russell W, Burch R (1959) The principle of humane experimental technique. Methuen, London

Russell W (2005) The three Rs: past, present, and future. Three Rs special issue, Animal welfare, vol 14, issue 4. Universities Federation for Animal Welfare, Wheathampstead, pp 279–286
Ryder R (1975) Victims of science – the use of animals in research. Davis-Poynter, London
Ryder R (1989) Animal revolution – changing attitudes towards speciesism. Blackwell, Oxford
Ryder R (1998) The political animal – the conquest of speciesism. McFarland, Jefferson, NC
Ryder R (2000) Animal revolution – changing attitudes towards speciesism. Berg, Oxford, Revised edition
Sachs J (2005) The end of poverty – economic possibilities for our time. Penguin, New York
Sankoff P (2008) Charting the growth of animal law in education. J Anim Law IV:105–148
Sankoff P, White S (eds) (2009) Animal law in Australasia – a new dialogue. Federation Press, Sydney
Sharman K (2007) From label to liable: scams, scandals and secrecy – lifting the veil on animal-derived food product labelling in Australia. A report prepared by Voiceless, Paddington, NSW, May 2007
Sharman K, Kossew S (2008) From nest to nugget – an exposé of Australia's chicken factories. A report prepared by Voiceless, Paddington, NSW, November 2008
Sherman B, Sherman O, Sharman K (2005) From paddocks to prisons – pigs in New South Wales current practices, future directions. A report prepared by Voiceless, Paddington, NSW, December 2005
Sibley R (2009) Book review of 'The future of animal farming: renewing the ancient contract' edited by M.S. Dawkins and R. Bonney (2008), Animal welfare. Universities Federation for Animal Welfare, Wheathampstead, pp 578–579
Singer P (1973) Animal liberation. Review of animals, men and morals edited by Stanley Godlovitch, Roslind Godlovitch and John Harris. The New York Review of Books, vol 20, no 5, April 5
Singer P (1975) Animal liberation – towards an end to man's inhumanity to animals. Jonathan Cape & Granada, London
Singer P (ed) (1985) In defence of animals. Blackwell, Oxford
Singer P (1990) Animal liberation, 2nd edn. Jonathan Cape, London
Singer P (1998) Ethics into action – Henry Spira and the animal rights movement. Rowman & Littlefield, Lanham, MD
Singer P (1999) The apes are trying to tell us something important. Anim Today 7(2):10
Singer P (2002) Animal liberation. Harper Collins, New York, pp ix–xiv, Preface to the 2002 edition
Singer P (2006a) Introduction. In: Singer P (ed) In defence of animals – the second wave. Blackwell, Oxford, pp 1–10
Singer P (ed) (2006b) Defense of animals – the second wave. Blackwell, Oxford
Singer P (2008) Foreword. In: Dawkins M, Bonney R (eds) The future of animal farming: renewing the ancient contract. Blackwell, Oxford, pp vii–ix
Singer P, Mason J (2006) The ethics of what we eat. Text Publishing, Melbourne
Singh S (ed) (2009) Who's who in Australia 2010. Crown Content, Melbourne
Smith A, Allen T (2005) The use of databases, information centres and guidelines when planning research that may involve animals. Three Rs special issue, Animal welfare, vol 14, issue 4. Universities Federation for Animal Welfare, Wheathampstead, pp 347–359
Spira H, Singer P (2006) Ten points for activists. In: Singer P (ed) In defense of animals – the second wave. Blackwell, Oxford, pp 214–224
Stafford K (2006) The welfare of dogs. Springer, Dordrecht
Steiner H, Alston P (2000) International human rights in context – law, politics, morals, text and materials, 2nd edn. Oxford University Press, Oxford
Stevenson P (2002) WTO: The greatest threat facing animal protection today. Compassion in World Farming Trust, Petersfield
Stoner L (2000) Farm animals. Anim Today 8(1):8–11

Sunstein C, Nussbaum M (eds) (2004) Animal rights – current debates and new directions. Oxford University Press, New York
The Advertiser (2006) Kangaroo birth pill. The Advertiser, Adelaide, 24 Aug 2006, p 2
Thornton P (2009) Book review of 'The animal ethics reader', first edition, edited by S.J. Armstrong and R.G. Botzler (2003), Animal welfare. Universities Federation for Animal Welfare, Wheathampstead, pp 110–111
Townend C (1981) A voice for the animals – how animal liberation grew in Australia. Kangaroo Press, Kenthurst, NSW
Townend C (1999) If the animals could say thanks ...: a tribute to Peter Singer. Anim Today 7(2):8–9
Trigwell S (ed) (2007) Animal experiments reach a 15-year high. FRAME News, no 62, October 2007, pp 10–11
Vallat B (2005) Preface. In: Bayvel ACD, Rahman SA, Gavinelli A (eds) Animal welfare: global issues, trends and challenges. World Organisation for Animal Health, Paris
Waran N (ed) (2002) The welfare of horses. Springer, Dordrecht
Webster J (1995) Animal welfare: a cool eye towards Eden. Blackwell, Oxford
Webster J (2005) Animal welfare: limping towards Eden. Blackwell, Oxford
Webster J (2006) Book review of 'In defence of animals: the second wave' edited by P. Singer (2005), Animal welfare. Universities Federation for Animal Welfare, Wheathampstead, pp 192–193
White S (2007) Animal law: how, or will it help to bring about a compassionate society? In: Humane education symposium, Griffith University, Brisbane, 5–6 Oct 2007
Wirth H (2005) Personal communication sent to the author by Dr Hugh Wirth AM KSJ, 25 Nov 2005
Wirth H (2006) Production animal welfare in the 21st Century. National Press Club, Canberra, 5 July 2006
WorldCat (2011) WorldCat, http://www.worldcat.org
World Organisation for Animal Health (2005) Animal welfare: global issues, trends and challenges. World Organisation for Animal Health, Paris
World Society for the Protection of Animals (2005a) Proposed Universal Declaration on Animal Welfare – Briefing, April 2005, WSPA, London
World Society for the Protection of Animals (2005b) Animals International, issue 70, Summer 05, WSPA

Index

DOI 10.1007/978-3-642-30577-1, © Springer-Verlag Berlin Heidelberg 2012

B

Q

R

S

The manufacturer's authorised representative in the EU is Springer Nature Customer Service Centre GmbH, Europaplatz 3, 69115 Heidelberg, Germany. If you have any concerns regarding our products, please contact ProductSafety@springernature.com

Printed and bound by CPI Group (UK) Ltd, Croydon, CR0 4YY
30/09/2026
02225472-0016